T0257236

EVERYDAY CALCULUS

EVERYDAY CALCULUS

Discovering the Hidden Math All around Us

OSCAR E. FERNANDEZ

PRINCETON UNIVERSITY PRESS
PRINCETON AND OXFORD

Copyright © 2014 by Princeton University Press
Published by Princeton University Press, 41 William Street,
Princeton, New Jersey 08540
In the United Kingdom: Princeton University Press,
6 Oxford Street, Woodstock, Oxfordshire OX20 1TW

press.princeton.edu

All Rights Reserved

Library of Congress Cataloging-in-Publication Data

Fernandez, Oscar E. (Oscar Edward)
Everyday calculus : discovering the hidden math all around us / Oscar E. Fernandez.
pages cm
Includes bibliographical references and index.
ISBN 978-0-691-15755-9 (hardcover : acid-free paper)
1. Calculus–Popular works.   I. Title.

QA303.2.F47 2014 515—dc232013033097

British Library Cataloging-in-Publication Data is available

This book has been composed in Minion Pro

Printed on acid-free paper. ∞

Typeset by S R Nova Pvt Ltd, Bangalore, India
Printed in the United States of America

1 3 5 7 9 10 8 6 4 2

Dedicado a Zoraida,
eres la belleza de mi vida
y también a nuestra hija
mi niña, tú serás mi consentida
y por supuesto a mi mamá
sin tu amor aquí no estuviera

CONTENTS

SINCE THE LATE 1600S, when calculus was being developed by the greatest mathematical minds of the day, scores of people across the world have asked the same question: When am I ever going to use this?

If you're reading this, you're probably interested in the answer to this question, as I was when I first started learning calculus. There are answers, like "Calculus is used by engineers when designing X," but this is more a statement of fact than an answer to the question. The pages that follow answer this question in a very different way, by instead revealing the hidden mathematics—calculus in particular—that describes our world.

To tell this revelatory tale I'll take you through a typical day in my life. You might be thinking: "A *typical* day? You're a mathematician! How typical can that be?" But as you'll discover, my day is just as normal as anyone else's. In the morning I sometimes feel groggy; I spend what feel like hours in traffic (even though they're only minutes) on my way to work; throughout my day I choose what to eat and where to eat it; and at some point I think about money. We don't pay attention to these everyday events, but in this book I'll peel back the facade of daily life and uncover its mathematical DNA.

Calculus will explain why our blood vessels branch off at certain angles (Chapter 5), and why every object thrown in the air arcs in the shape of a parabola (Chapter 1). Its insights will make us rethink what we know about time and space, demonstrating that we can time travel into the future (Chapter 3), and that our universe is expanding (Chapter 7). We'll also see how calculus can help us awake feeling more rested (Chapter 1), cut down on our car's fuel consumption (Chapter 5), and find the best seat in a movie theater (Chapter 7).

So, if you've ever wondered what calculus can be used for, you should have a hard time figuring out what it *can't* be used for after reading this book. The applications we'll discuss will be accompanied throughout the chapters by various formulas. These equations will gently help you build your mathematical understanding of calculus, but don't worry if you're a bit rusty with your math; you won't need to understand any of them to enjoy the book. But in case you're curious about the math, Appendix A includes a refresher on functions and graphs to get you started, and appendices 1–7 include the calculations mentioned throughout the book, which are indicated by superscripts that look like this.[*1] (You'll also find footnotes indicated by Roman numerals and endnotes indicated by Arabic numerals.) Finally, on the next page you'll find a breakdown of the mathematics discussed in each chapter.

Whether you're new to calculus, you're studying calculus, or it's been a few years since you've seen it, you'll find a whole new way of looking at the world in the next few chapters. You may not see fancy formulas flashing before your eyes when you finish this book, but I'm hopeful that you'll achieve an enlightenment akin to what Neo in *The Matrix* experiences when he learns that a computer code underlies his reality. Although I'm not as cool as Morpheus, I look forward to helping you emerge through the other end of the rabbit hole.

Oscar Edward Fernandez
Newton, MA

CALCULUS TOPICS DISCUSSED BY CHAPTER

The chart below details the calculus topics discussed in each chapter.

Chapter 1
                              Linear Functions
Polynomial Functions
Trigonometric Functions
Exponential Functions
Logarithmic Functions

Chapter 2
                              Slopes and Rates of Change
Limits and Derivatives
Continuity

Chapter 3
                              Interpreting the Derivative
The Second Derivative
Linear Approximation

Chapter 4
                              Differentiation Rules
Related Rates

Chapter 5
                              Differentials
Optimization
The Mean Value Theorem

Chapter 6
                              Riemann Sums
Area under a Curve
The Definite Integral
The Fundamental Theorem of Calculus
Antiderivatives
Application of Integration to Wait Times

Chapter 7
                              Average Value of a Function
Arc Length of a Curve
Application to the Best Theater Seat
Application to the Age of the Universe

EVERYDAY CALCULUS

CHAPTER 1

WAKE UP AND SMELL THE FUNCTIONS

IT's FRIDAY MORNING. The alarm clock next to me reads 6:55 a.m. In five minutes it'll wake me up, and I'll awake refreshed after sleeping roughly 7.5 hours. Echoing the followers of the ancient mathematician Pythagoras—whose dictum was "All is number"—I deliberately chose to sleep for 7.5 hours. But truth be told, I didn't have much of a choice. It turns out that a handful of numbers, including 7.5, rule over our lives every day. Allow me to explain.

A long time ago at a university far, far away I was walking up the stairs of my college dorm to my room. I lived on the second floor at the time, just down the hall from my friend Eric Johnson's room. EJ and I were in freshman physics together, and I often stopped by his room to discuss the class. This time, however, he wasn't there. I thought nothing of it and kept walking down the narrow hallway toward my room. Out of nowhere EJ appeared, holding a yellow Post-it note in his hand. "These numbers will change your life," he said in a stern voice as he handed me the note. Off in the corner was a sequence of numbers:

$$1.5 \quad 4.5 \quad 7.5$$

$$3 \quad 6$$

Like Hurley from the *Lost* television series encountering his mystical sequence of numbers for the first time, my gut told me that these numbers meant something, but I didn't know what. Not knowing how to respond, I just said, "Huh?"

EJ took the note from me and pointed to the number 1.5. "One and a half hours; then another one and a half makes three," he said. He explained that the average human sleep cycle is 90 minutes (1.5 hours) long. I started connecting the numbers in the shape of a "W." They were all a distance of 1.5 from each other—the length of the sleep cycle. This was starting to sound like a good explanation for why some days I'd wake up "feeling like a million bucks," while other days I was just "out of it" the entire morning. The notion that a simple sequence of numbers could affect me this much was fascinating.

In reality getting *exactly* 7.5 hours of sleep is very hard to do. What if you manage to sleep for only 7 hours, or 6.5? How awake will you feel then? We could answer these questions if we had the sleep cycle *function*. Let's create this based on the available data.

What's Trig Got to Do with Your Morning?

A typical sleep cycle begins with REM sleep—where dreaming generally occurs—and then progresses into non-REM sleep. Throughout the four stages of non-REM sleep our bodies repair themselves,[1] with the last two stages—stages 3 and 4—corresponding to deep sleep. As we emerge from deep sleep we climb back up the stages to REM sleep, with the full cycle lasting on average 1.5 hours. If we plotted the sleep stage S against the hours of sleep t, we'd obtain the diagram in Figure 1.1(a). The shape of this plot provides a clue as to what function we should use to describe the sleep stage. Since the graph repeats roughly every 1.5 hours, let's approximate it by a *trigonometric function*.

To find the function, let's begin by noting that S depends on how many hours t you've been sleeping. Mathematically, we say that your sleep stage S is a function of the number of hours t you've been asleep,[1] and write $S = f(t)$. We can now use what we know about sleep cycles to come up with a reasonable formula for $f(t)$.

---

[1] Appendix A includes a short refresher on functions and graphs.

Since we know that our REM/non-REM stages *cycle* every 1.5 hours, this tells us that $f(t)$ is a *periodic function*—a function whose values repeat after an interval of time T called the *period*—and that the period $T = 1.5$ hours. Let's assign the "awake" sleep stage to $S = 0$, and assign each subsequent stage to the next negative whole number; for example, sleep stage 1 will be assigned to $S = -1$, and so on. Assuming that $t = 0$ is when you fell asleep, the trigonometric function that results is [*1]

$$f(t) = 2\cos\left(\frac{4\pi}{3}t\right) - 2,$$

where $\pi \approx 3.14$.

Before we go off and claim that $f(t)$ is a good mathematical model for our sleep cycle, it needs to pass a few basic tests. First, $f(t)$ should tell us that we're awake (sleep stage 0) every 1.5 hours. Indeed, $f(1.5)=0$ and so on for multiples of 1.5. Next, our model should reproduce the actual sleep cycle in Figure 1.1(a). Figure 1.1(b) shows the graph of $f(t)$, and as we can see it does a good job of capturing not only the awake stages but also the deep sleep times (the troughs).[ii]

In my case, though I've done my best to get exactly 7.5 hours of sleep, chances are I've missed the mark by at least a few minutes. If I'm way off I'll wake up in stage 3 or 4 and feel groggy; so I'd like to know how close to a multiple of 1.5 hours I need to wake up so that I still feel relatively awake.

We can now answer this question with our $f(t)$ function. For example, since stage 1 sleep is still relatively light sleeping, we can ask for all of the t values for which $f(t) \geq -1$, or

$$2\cos\left(\frac{4\pi}{3}t\right) - 2 \geq -1.$$

The quick way to find these intervals is to draw a horizontal line at sleep stage -1 on Figure 1(b). Then all of the t-values for which our graph is

---

[ii]As Figure 1.1(a) shows, after roughly three full sleep cycles (4.5 hours of sleep) we don't experience the deep sleep stages again. We didn't factor this in when designing the model, which explains why $f(t)$ doesn't capture the shallower troughs seen in Figure 1.1(a) for $t > 5$.

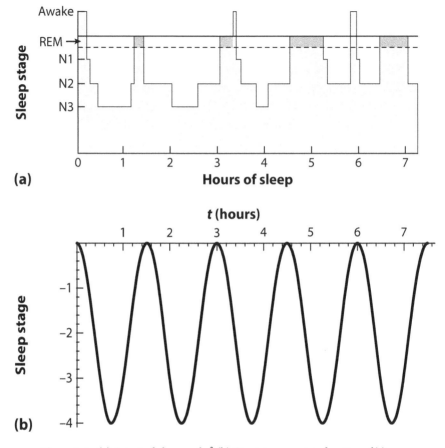

Figure 1.1. (a) A typical sleep cycle.[2] (b) Our trigonometric function $f(t)$.

above this line will satisfy our inequality. We could use a ruler to obtain good estimates, but we can also find the exact intervals by solving the equation $f(t) = -1$:[*2]

$$[0, 0.25], \quad [1.25, 1.75], \quad [2.75, 3.25], \quad [4.25, 4.75], \quad [5.75, 6.25],$$
$$[7.25, 7.75], \quad \text{etc.}$$

We can see that the endpoints of each interval are 0.25 hour—or 15 minutes—away from a multiple of 1.5. Hence, our model shows that missing the 1.5 hour target by 15 minutes on either side won't noticeably impact our morning mood.

This analysis assumed that 90 minutes represented the *average* sleep cycle length, meaning that for some of us the length is closer to 80 minutes, while for others it's closer to 100. These variations are easy to incorporate into $f(t)$: just change the period T. We could also replace the 15-minute buffer with any other amount of time. These *free parameters* can be specified for each individual, making our $f(t)$ function very customizable.

I'm barely awake and already mathematics has made it into my day. Not only has it enabled us to solve the mystery of EJ's multiples of 1.5, but it's also revealed that we all wake up with a built-in trigonometric function that sets the tone for our morning.

How a Rational Function Defeated Thomas Edison, and Why Induction Powers the World

Like most people I wake up to an alarm, but unlike most people I set *two* alarms: one on my radio alarm clock plugged into the wall and one on my iPhone. I adopted this two-alarm system back in college when a power outage made me late for a final exam. We all know that our gadgets run on electricity, so the power outage must have interrupted the flow of electricity to my alarm clock at the time. But what is "electricity," and what causes it to flow?

On a normal day my alarm clock gets its electricity in the form of *alternating current* (AC). But this wasn't always the case. In 1882 a well-known inventor—Thomas Edison—established the first electric utility company; it operated using *direct current* (DC).[3] Edison's business soon expanded, and DC current began to power the world. But in 1891 Edison's dreams of a DC empire were crushed, not by corporate interests, lobbyists, or environmentalists, but instead by a most unusual suspect: a *rational function*.

The story of this rational function begins with the French physicist André-Marie Ampère. In 1820 he discovered that two wires carrying electric currents can attract or repel each other, as if they were magnets. The hunt was on to figure out how the forces of electricity and magnetism were related.

The unexpected genius who contributed most to the effort was the English physicist Michael Faraday. Faraday, who had almost no formal education or mathematical training, was able to visualize the interactions between magnets. To everyone else the fact that the "north" pole of one magnet attracted the "south" pole of another—place them close to each other and they'll snap together—was just this, a fact. But to Faraday there was a *cause* for this. He believed that magnets had "lines of force" that emanated from their north poles and converged on their south poles. He called these lines of force a *magnetic field*.

To Faraday, Ampère's discovery hinted that magnetic fields and electric current were related. In 1831 he found out how. Faraday discovered that moving a magnet near a circuit creates an electric current in the circuit. Put another way, this *law of induction* states that a changing magnetic field produces a *voltage* in the circuit. We're familiar with voltages produced by batteries (like the one in my iPhone), where chemical reactions release energy that results in a voltage between the positive and negative terminals of the battery. But Faraday's discovery tells us that we don't need the chemical reactions; just wave a magnet near a circuit and voilà, you'll produce a voltage! This voltage will then push around the electrons in the circuit, causing a flow of electrons, or what we today call *electricity* or *electric current*.

So what does Edison have to do with all of this? Well, remember that Edison's plants operated on DC current, the same current produced by today's batteries. And just like these batteries operate at a fixed voltage (a 12-volt battery will never magically turn into a 15-volt battery), Edison's DC-current plants operated at a fixed voltage. This seemed a good idea at the time, but it turned out to be an epic failure. The reason: hidden mathematics.

Suppose that Edison's plants produce an amount V of electrical energy (i.e., voltage) and transmit the resulting electric current across a power line to a nineteenth-century home, where an appliance (perhaps a fancy new electric stove) sucks up the energy at the constant rate P_0. The radius r and length l of the power line are related to V by

$$r(V) = k\frac{\sqrt{P_0 l}}{V},$$

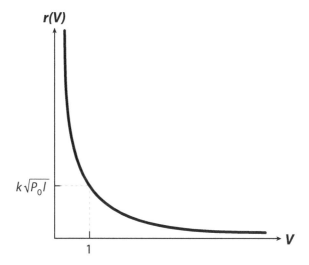

Figure 1.2. A plot of the rational function $r(V)$.

where k is a number that measures how easily the power line allows current to flow.[iii] This *rational function* is the nemesis Edison never saw coming.

For starters, the easiest way to distribute electricity is through hanging power lines. And there's an inherent incentive to make these as thin (small r) as possible, otherwise they would both cost more and weigh more—a potential danger to anyone walking under them. But our rational function tells us that to carry electricity over large distances (large l) we need large voltages (large V) if we want the power line radius r to be small (Figure 1.2). And this was precisely Edison's problem; his power plants operated at the low voltage of 110 volts. The result: customers needed to live at most 2 miles from the generating plant to receive electricity. Since start-up costs to build new power plants were too high, this approach soon became uneconomical for Edison. On top of this, in 1891 an AC current was generated and transported *108 miles* at an exhibition in Germany. As they say in the sports business, Edison bet on the wrong horse.[4]

[iii] This property of a material is called the *electrical resistivity*. Power lines are typically made of copper, since this metal has low electrical resistivity.

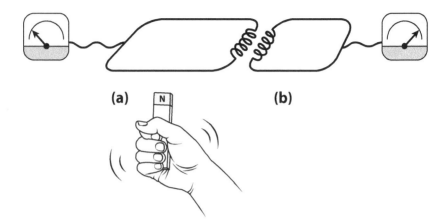

Figure 1.3. Faraday's law of induction. (a) A changing magnetic field produces a voltage in a circuit. (b) The alternating current produced creates another changing magnetic field, producing another voltage in a nearby circuit.

But the function $r(V)$ has a split personality. Seen from a different perspective, it says that if we crank up the voltage V—by a lot—we can also increase the length l—by a bit less—and *still* reduce the wire radius r. In other words, we can transmit a very high voltage V across a very long distance l by using a very thin power line. Sounds great! But having accomplished this we'd still need a way to transform this high voltage into the low voltages that our appliances use. Unfortunately $r(V)$ doesn't tell us how to do this. But one man already knew how: our English genius Michael Faraday.

Faraday used what we mathematicians would call "transitive reasoning," the deduction that if A causes B and B causes C, then A must also cause C. Specifically, since a changing magnetic field produces a current in a circuit (his law of induction), and currents flowing through circuits produce magnetic fields (Ampère's discovery), then it should be possible to use magnetic fields to *transfer* current from one circuit to another. Here's how he did it.

Picture Faraday—a clean-shaven tall man with his hair parted down the middle—with a magnet in his hand, waving it around a nearby circuit. Induction causes this changing magnetic field to produce a voltage V_a in one circuit (Figure 1.3(a)). The alternating current produced would, by Ampère's discovery, produce another changing magnetic field. The result would be *another* voltage V_b in a *nearby* circuit (Figure 1.3(b)), producing current in that circuit.

As Faraday waves the magnet around, sometimes he does so closer to the loop and sometimes farther away; sometimes he waves it fast and other times slow. In other words, the voltage V_a produced *changes*. Today, magnets are put inside objects like windmills that do the waving for us. As the blades rotate in the wind, the magnetic field produced inside the turbine also changes. In this case the changes are described by a *trigonometric function* (not by Faraday's crazy hand-waving). This alternating voltage causes the current to alternate too, putting the "alternating" in alternating current.

Great, we can now transfer current between circuits. But we still have the voltage problem: most household plugs run at low voltages (a fact left over from Edison's doings), yet our modern grids produce voltages as high as 765,000 volts; how do we reduce this to the standard range of 120–220 volts that most countries use?

Let's suppose that the original circuit's wiring has been coiled into N_a turns, and that the nearby circuit's wiring has been coiled into N_b turns (Figure 1.4(a)). Then

$$V_b = \frac{N_b}{N_a} V_a.$$

This formula says that a high incoming voltage V_a can be "stepped down" to a low outgoing voltage V_b by using a large number of turns N_a for the incoming coiling relative to the outgoing coiling. This transfer of voltage is called *mutual induction*, and is at the heart of modern electricity transmission. In fact, if you step outside right now and look up at the power lines you'll likely see cylindrical buckets like the one in Figure 1.4(b). These *transformers* use mutual induction to step down the high voltages produced by modern electricity plants to lower, safer voltages for household use.

The two devices that got me going on this story—the iPhone and my clock radio—honor the legacies of both Edison and Faraday. My iPhone runs on DC current from its battery, and my clock radio draws

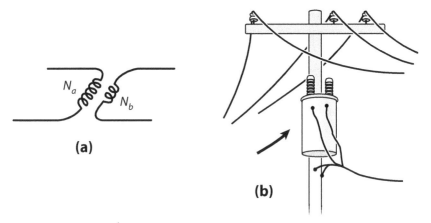

Figure 1.4. (a) Two circuits with different coiling numbers N_a and N_b. (b) A drawing of a transformer.

its power from the AC current coming through the wall plug, itself produced *dozens* of miles away at the electricity plant by an alternating voltage. And somewhere in between, Faraday's mutual induction is at work stepping down the voltage so that we can power our devices.

But the real hero here is the rational function $r(V)$. It spelled doom for Edison, but through a differnt interpretation suggested that we base our electric grid on voltages *much* higher than Edison's 110 volts. This idea of "listening" closely to mathematics to learn more about our world is a recurring theme of this book. We've already exposed two functions—the trigonometric $f(t)$ and the rational $r(V)$—that follow you around everywhere you go. Let me wake up so that I can reveal even more hidden mathematics.

The Logarithms Hidden in the Air

It's now seven in the morning and my alarm clock finally goes off. It's set to play the radio when the alarm goes off, rather than that startling "BUZZ! BUZZ!" I can't stand. Back when I lived in Ann Arbor I would wake up to 91.7 FM, the local National Public Radio (NPR) station.

But now that I live in Boston, 91.7 FM is pure static. What happened to the Ann Arbor station? Is my radio broken? Where's my NPR?!

The local NPR station for Boston is WBUR-FM, at 90.9 FM on the radio dial. Since I'm now far away from Ann Arbor my radio can't pick up the old 91.7 NPR station. We all intuitively know this; just drive far enough away from your home town and all your favorite radio stations will fade away. But wait a second, that's the same relationship that we saw in Figure 1.2 with the function $r(V)$. Could there be another rational function lurking somewhere in the air waves?

Let's get back to WBUR to figure this out. The station's "effective radiative power"—a measure of its signal strength—is 12,000 watts.[5] You should recognize the unit here from your experience with light bulbs; just as a 100-watt bulb left on for one hour would *consume* 100 watt-hours of energy, WBUR's station *emits* 12,000 watt-hours of energy in one hour. That's the equivalent of 12,000/100 = 120 light bulbs worth of energy every hour! But where does that energy go?

Picture a light bulb placed on the floor in the middle of a dark room. Turn it on and the light it emits will light up everything in the room. The bulb radiates its energy, partly in the form of light, evenly throughout the space in the room. Similarly, WBUR's antenna radiates its energy outward in the form of *radio waves*.

Now, just as you'd perceive the bulb's light to be brighter the closer you are to it, the radio signals coming from WBUR's antenna come through clearer when you're closer to the antenna. We can measure this by calculating the *intensity* $J(r)$ of the signal at a distance r from the antenna:

$$J(r) = \frac{\text{radiated power}}{\text{surface area}} = \frac{12{,}000}{4\pi r^2} = \frac{3{,}000}{\pi r^2}, \qquad (1)$$

where I've assumed that the energy is radiated spherically outward. Aha! Here's the *rational* function we had predicted. Let's see if we can "listen" to it and learn something about how radios work.

The $J(r)$ formula tells us that the intensity of the signal decreases as the distance r from the antenna increases. This explains what happened in my move from Ann Arbor to Boston: it's not that the Ann Arbor

NPR station doesn't reach me anymore, but that its signal intensity is too weak to be picked up by my radio. On the other hand, at my current distance from WBUR's antenna my radio has no problem picking up the station.

While I lie there still in a haze, I pick up a few headlines from the voice coming out of the radio; something about the economy and later about politics. Nothing too exciting so I just stay in bed, listening. There's always a danger I'll fall back asleep (think "second alarm"); to thwart this I decide to boot up my brain by asking a simple question: what am I listening to?

Certainly the answer is WBUR at 90.9 FM. But that's a radio wave, and we humans can't *hear* a radio wave; the ear's frequency range is from 20 to 20,000 hertz,[6] and WBUR's signal is broadcast at 90.9 megahertz.[iv] Ergo, it's not the radio wave I hear. What I hear is the *sound* waves coming from my radio. And somehow that little gadget manages to convert a radio wave—which I can't hear—into a sound wave, which I can. But how?

Part of the answer is hidden in the fact that WBUR transmits at 90.9 megahertz. All sounds have a frequency associated with them; for example, the 49th key—called A4—on an 88-key piano has a frequency of 440 hertz. And we know (either from Appendix A or from your general knowledge) that phenomena with frequencies can be represented as oscillating functions, just like our sleep cycle functions. But then, what's oscillating in this case? *Something* has to move back and forth between the radio and my ear. And the only possibility is air, so the answer must be related to changes in *air pressure.*

In a nutshell, *sound is a pressure wave.* This is easy to confirm: hold your palm very close to your mouth and try to speak without any air hitting your palm. Good luck, because without the movement of air molecules there's no pressure wave. Now hold your hand somewhat close to your ear and fan it ferociously back and forth. You should hear a periodic sound as your arm oscillates: that sound is the pressure wave.

[iv]1 megahertz (MHz) is 1×10^6 hertz. Hertz (Hz) is the unit of frequency (see Appendix A for a quick refresher).

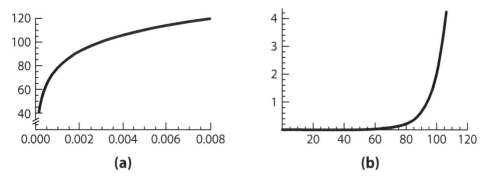

Figure 1.5. (a) A plot of the function $L(p)$. (b) A plot of the function $p(L)$.

Like your arm, a radio pulses its speakers back and forth to produce the pressure waves that our ears detect as sound. And just like your arm, the more violently the speakers vibrate the louder the sound that's created. Mathematically, if we denote by p the sound pressure of a pressure wave, then the "sound level" $L(p)$ of that sound is given by the *logarithmic* function (Figure 1.5(a))

$$L(p) = 20 \log_{10}(50{,}000\,p) \text{ decibels.}$$

Let's examine the familiar *decibel* (dB) units. As a reference, the water coming out of a showerhead makes a sound of about 80 decibels, and a jet engine at about 100 feet makes a sound of 140 decibels. From these numbers you can see why long-term exposure to sounds at levels as low as 90 decibels has the potential to cause hearing loss.[7] We're all more used to the decibel scale than to measuring pressure waves, so lets invert the $L(p)$ equation. We arrive at the exponential function[*3] (Figure 1.5(b))

$$p(L) = \frac{1}{50{,}000} 10^{L/20}.$$

The $p(L)$ equation tells us that, for example, a sound level of $L = 0$ decibels gives a pressure of $p(0) = 1/50{,}000 = 20 \times 10^{-6}$ pascals, the unit of pressure. This sound level and pressure combination roughly correspond to the sound a mosquito would make as it flaps its wings roughly 10 feet away from you,[8] hence the small pressure number.

Now that I've gotten myself up and about figuring out this pressure thing, a nagging thought has developed in my head. Just a few minutes ago I was somewhere along my sleep cycle—modeled by trigonometric function $f(t)$—and then my radio turned on, thanks to our rational function $r(V)$ and WBUR's antenna intensity function $I(r)$. The NPR reporter's voice then created a pressure wave that I interpreted as sound via the $L(p)$ function (we actually *hear* logarithmic functions; how cool is that?). There's so much going on. Is there any order to this chaos? Does my morning consist of chance encounters with different functions, or are they all related somehow? A hierarchy or a unifying principle would be nice.

The Frequency of Trig Functions

My new quest gives me something to think about while I pick my clothes out. On the other end of the bedroom is a small closet that my wife Zoraida and I cram our clothes into. I'm shuffling clothes around looking for something to wear after I shower. In the background, a soft sound begins to steadily increase in intensity; Zoraida is snoring. I figure I'll wake her up (we've got to get to work soon) by turning on the TV; she likes waking up to the morning shows. Naturally, I reach for another one of our modern gadgets: the remote control.

With the control in hand, I push the "channel up" button, looking for something she'd like. The remote sends out *infrared light waves* at frequencies of about 36,000 hertz. Although I can't see these signals— they are outside our frequency range of vision—the pulses of 1's and 0's that are emitted instruct the TV to change to the next channel. I find one of those morning shows and put the volume just loud enough to eventually wake her up.

Now that I've picked out a pair of khakis and a shirt, I get back to thinking about this "unifying principle" business on my way to the shower. The hallway's dark; it's a cloudy day outside. I'm hoping that since it's July the rain will quickly be followed by sunshine. This triggers a memory of a conversation I had back in high school with my friend Blake about light. We were talking about how the colors we see are

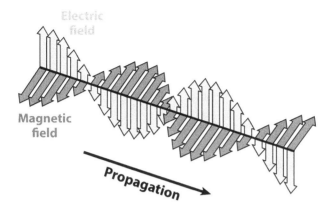

Figure 1.6. An electromagnetic wave. The electric and magnetic fields it carries oscillate perpendicular to each other as the wave propagates. Image from http://www.molphys.leidenuniv.nl/monos/smo/index.html?basics/light.htm.

described by different frequencies of light. For example, red light has a frequency range of about 430 to 480 terahertz.[v,9] Blake was wondering if aliens would see red light—light in the frequency range of 430 to 480 terahertz—as actually "red." This was in biology class, so we spent some time talking about what our eyes think "red" is.

Midway through my recollection I'm interrupted by a simple, clearly articulated word: frequency. And then it clicks. The AC current, the radio waves, the infrared waves, and sunlight, they all have a frequency associated with them. Here's the unifying principle I've been looking for! Because they are characterized by a frequency, these are all oscillating functions—*trigonometric* functions.

This *mathematical* unifying principle also has a *physical* analogue. All of these waves—with the exception of AC current, which we'll discuss shortly—are all particular types of *electromagnetic waves* (EM waves for short). As the name suggests, an electromagnetic wave carries along with it an electric *and* a magnetic field.[vi] These fields oscillate perpendicular to each other as the wave propagates, and each can be represented by trigonometric functions (Figure 1.6).

[v]One terahertz (THz) is 1×10^{12} hertz.
[vi]An electric field is the analogue of a magnetic field, where positive and negative charges play the roles of the north and south poles of a magnet.

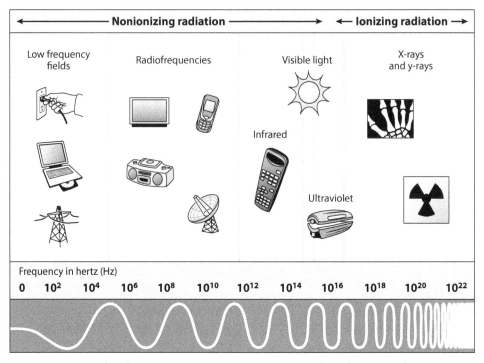

Figure 1.7. The electromagnetic wave spectrum. Image from http://www.hermes-program.gr/en/emr.aspx.

One of the greatest discoveries of the nineteenth century—due to our induction discoverer Michael Faraday—was that light itself is an EM wave. This explains why light has a *frequency* associated with it. Thus, infrared light, radio waves, and any other radiation that has a frequency associated with it is an EM wave (Figure 1.7). Alternating current, although not an electromagnetic wave itself, emits electromagnetic waves as it travels down a wire. An electromagnetic wave, along with its mathematical representation as a trigonometric function, is the unifying concept I was looking for.

When I turn on the light in my bathroom, I pause for a second to marvel at all the EM waves around me. The light the bulb produces? An EM wave. The sunlight coming through the window? Another EM wave. The radio waves transmitting NPR to the bedroom radio? Yep, just another electromagnetic wave. So, not only can we *hear* logarithms (recall our function $L(p)$); now you know that we can *see* trigonometric functions (light). Who knew that trigonometric functions occurred so frequently throughout the day? (Pun intended.)

Galileo's Parabolic Thinking

I turn the tub's faucet on and switch on the showerhead; the water is freezing! It'll take a minute or so for it to heat up. No problem, I'll just brush my teeth while I wait. While I brush up and down, left and right (don't worry, I won't mention the trigonometric function here; oops, I just did!), I continue looking at the water stream, as if that'll make it heat up faster.

Inspired by Faraday's ability to see magnetic fields, I start trying to see the "gravitational field" and its effect on the stream. I know the field exists, since the water doesn't shoot out in a straight line, even though it comes out of the showerhead with a high velocity; instead it looks like it's "attracted" to the floor. Of course, there's no magnetism going on here, it's just gravity, but that's the *physics*. What about the *math*? The man who figured this out, Galileo Galilei, was referred to by none other than Einstein himself as the "father of modern science." He built powerful telescopes, and later used it to decisively confirm that the Earth revolves around the Sun and not the other way around. In addition, Galileo is also well known for his experiments with falling objects. The most famous of these is the Leaning Tower of Pisa experiment. Vincenzo Viviani, Galileo's pupil, described the experiment in a biography of Galileo. He wrote that Galileo had dropped balls of different masses from the tower to test the conjecture that they would reach the ground at the same time, regardless of their mass.[vii, 10] Galileo, in his earlier writings, had proposed that a falling object would fall with a uniform (constant) acceleration. By using this simple proposition, he had also demonstrated mathematically that the

[vii] This popular story might actually be a legend, but Vincenzo is no longer around to set the record straight.

(a)

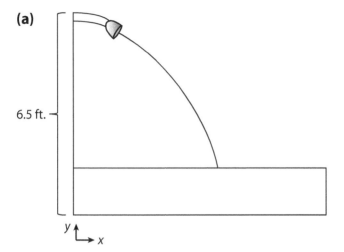

6.5 ft.

(b)

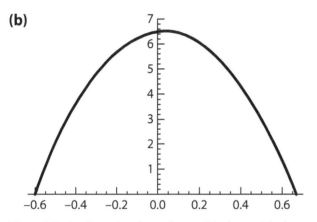

Figure 1.8. A schematic of my shower (a) along with the graph of the parabolic function $y(x) = 6.5 + x - 16x^2$ (b).

distance the object traveled would be proportional to the square of the amount of time the object was in motion.[11]

To fully appreciate this result, let's consider what it means in the context of the water coming out of my showerhead. Figure 1.8(a) shows a profile of my shower. We can define a *coordinate system* whose origin is on the ground, directly underneath my showerhead. Let's call the horizontal direction x and the vertical direction y, and suppose that the water is coming out of the showerhead with a constant speed of v_x

in the x-direction and v_y in the y-direction. Since gravity acts only in the vertical direction, *there is no acceleration in the horizontal direction* (as the joke goes, "sometimes gravity gets me down," but never "left," "right," or "up"). We can now use the familiar formula *distance* $=$ *rate* $\times$ *time* to determine the *horizontal* distance $x(t)$ traveled by a water molecule:

$$x(t) = v_x t,$$

where we'll measure t in seconds since the water molecule left the showerhead.

What about the vertical (y) direction motion? Each water molecule coming out of the showerhead is being pulled down by gravity, which Galileo says accelerates objects at a constant rate; let's denote this by $-g$, where the negative sign is there to remind us that this acceleration is downward. Using this, along with the fact that our water molecule's initial speed is v_y at time $t = 0$ and what we will call $v(t)$ at time $t > 0$, we then find*[4] that our water molecule's vertical speed $v(t)$ is the *linear* function

$$v(t) = v_y - gt,$$

its initial speed plus the contribution from gravity. It was also known in Galileo's time that the distance traveled by objects whose speed varies linearly with time is given by

$$y(t) = y_0 + v_{avg}t, \qquad \text{where} \qquad v_{avg} = \frac{1}{2}\left(v_{initial} + v_{final}\right),$$

where y_0 is the initial position of the object. For our water molecule, since its vertical position is 6.5 feet above the ground when it comes out of the showerhead, we know that $y_0 = 6.5$. Moreover, since its initial vertical speed was v_y and its final vertical speed was $v(t) = v_y - gt$, then its average speed is

$$v_{avg} = v_y - \frac{1}{2}gt, \qquad \text{so that} \qquad y(t) = 6.5 + v_y t - \frac{1}{2}gt^2.$$

Unlike the $x(t)$ formula, the water molecule's vertical position is a *polynomial* function of t; more specifically, it's a *quadratic* function.

We can put these two formulas together by solving the $x(t)$ equation for t and substituting the result into the $y(t)$ formula. We arrive at[*5]

$$y(x) = 6.5 + \frac{v_y}{v_x}x - \frac{g}{2v_x^2}x^2.$$

Since v_x, v_y, and g are numbers, this formula can be put in the form $y = 6.5 + Bx - Ax^2$, which is the equation for a *parabola* (Figure 1.8(b)). And since the coefficient of x^2 is negative, this parabola opens *downward*. Therefore, the mathematics is telling us that the water coming out of my shower bends toward the ground. And that's exactly what happens!

This formula, in my opinion, is one of the greatest achievements of medieval science. It applies not just to the water coming out of my showerhead, but also to a football, a Frisbee, or any other object thrown in the air. It tells us that *all objects (of reasonable mass) thrown upward on Earth follow parabolic trajectories*. To medieval scientists working at a time when religion was the predominant way to understand the world, results like these were seen as glimpses into the mind of God. They inspired future scientists to continue applying mathematics to our world in the hopes of achieving equally profound insights.

We'll spend the next chapter talking about one such scientist—Isaac Newton—who followed in Galileo's footsteps and made equally revolutionary advances for his time. For now, I hope this chapter has convinced you that functions are not abstract mathematical constructs. Instead, as Galileo and Faraday showed us, they can be seen, heard, and felt all around us every day. The journey that got us here started with the Pythagoreans' belief that "All is number," but this chapter suggests the more current Pythagorean-like dictum: "All are functions."

CHAPTER 2

‖‖‖

BREAKFAST AT NEWTON'S

EVERYONE HAS A MORNING ROUTINE. After my shower, I like to tune to the financial news network CNBC while I get dressed. Its morning show is the closest I can get to a daily TV show about mathematics.[viii] In five minutes of watching it you're likely to see changes in interest rates, rising and falling stock market prices, fluctuating currency exchange rates, and, well, lots of other numbers flashing red and green.

After years of starting my mornings like this, I'm used to this barrage of information. But not my wife, Zoraida; this particular channel gives her a headache. "There are numbers racing across the screen in all directions; there's just way too much stuff going on," she says. I agree. But to me, the fact that CNBC's abundance of *change* is expressed through numbers hints at deeper mathematics. If functions describe our world—as I tried to convince you in the previous chapter—what function describes how the world around us *changes*? Mathematicians spent almost *two millennia* searching for the answer, but don't worry, after this chapter you'll see this "change function" everywhere.

Introducing Calculus, the CNBC Way

On this particular morning, CNBC is abuzz with information on the computer giant Apple. The new iPhone will be launched soon, and the news anchors, in discussing the impact on the company's stock, flash up a graph of Apple's (AAPL) stock price (Figure 2.1).

[viii]Sure there's *Numbers*, *Fringe*, or even *The Big Bang Theory*, but CNBC is on *all day*.

Figure 2.1. AAPL's stock price for the year ending July 31, 2012. Retrieved from http://www.stockcharts.com, July 31, 2012.

The anchors state that over the past year the stock has been a winner, appreciating by roughly $221 per share. However, they point out it's down almost $25 since its peak in early April. In math-speak, the anchors are providing us with *average rates of change*.

To spot a rate, look at the *units* of the number. All rates, including these average rates of change, have units that are ratios of other units. For example, we measure speed in miles per hour. But sometimes, as is the case this morning, some of the units are hidden. Sure, Apple's share price is measured in dollars, but what's the other unit in the anchors' statements? Time. (The phrases "over the past year" and "since early April" are the tipoffs.)

But merely spotting a rate won't help us figure out what the "change function" we're looking for is. So let's slow down a bit and define precisely what an average rate of change is.

Mathematically, if we denote by t the number of months since July 31, 2011, and by $P(t)$ the price of AAPL, then the average rate of change (AROC) of the stock's price between months $t = a$ and $t = b$ is simply the change in price divided by the change in time:

$$m_{\text{avg}} = \frac{P(b) - P(a)}{b - a}. \tag{2}$$

Looking back at the price chart, we see that at the start of the chart ($t = 0$) AAPL was trading at about \$390, at $t = 8$ it was trading at about \$625, and at $t = 12$ it was trading at \$610.76. Using these values, we find that the stock's price appreciated roughly \$18.40 per month over the past year, while dropping roughly \$3.60 each month over the last four months.[*1]

These AROCs are useful information, no doubt, but I'm a visual person. For me, it's easier to understand AROCs if they're represented graphically. Another reporter reads my mind, and on his fancy touchscreen he draws a line on AAPL's chart. It begins on July 31, 2011, and ends on July 31, 2012. Now, if you recall what you learned about linear functions—or if you've already familiarized yourself with equation (94) in Appendix A—you'll recognize that calculating the slope of the line drawn by the reporter is the same as calculating the AROC!

Why the exclamation? It's because this revelation gives us a *geometric* way of calculating the AROC. Simply draw a line between any two points on the chart in Figure 2.1, find the slope of that line, and your answer is just the AROC between those two points. The line you obtain is often called the *secant line*.

By the time the reporter is done drawing on his screen, AAPL's chart looks like a page out of a football playbook. And although he's done a good job of describing how AAPL's price has changed over time, what you'll never hear him say is how the stock's price is changing *at this instant*. Here's why.

Mathematically, instants present a problem for our AROC formula (2). Due to the $b - a$ in the denominator, the formula works only for *intervals* of time (where $b - a \neq 0$), and not *instants* of time, where $b = a$. At an instant, the denominator becomes zero, and we can't divide a number by zero. So to describe AAPL's price at a particular *instant* of time what we really need to be able to compute is an *instantaneous rate of change* (IROC); here's how we do it.

Let's first pick a start for our time frame, say April 1, 2012 (which is $t = 8$), so that $a = 8$. Although we can't have a zero in the denominator of our AROC formula (2), we *can* make it as close to zero as we'd like by choosing b to be as close to 8 as we'd like.

To do this, let's denote by h the number of months *after* $t = 8$. For example, $h = 1$ corresponds to $t = 9$. Then formula (2) says that the *average* rate of change between $t = 8$ and $t = 8 + h$ is*[2]

$$m_{\text{avg}} = \frac{P(8 + h) - P(8)}{h}. \tag{3}$$

By choosing different (nonzero) values for h, this new formula will give us AROCs that describe how AAPL's price was changing between April 1, 2012, and some amount of time h afterwards. But what about the geometric interpretation?

Figure 2.2 shows AAPL's price (dashed curve) zoomed in to $t = 8$, along with two secant lines (the two thin lines) corresponding to $h = 0.5$ and $h = 0.1$. Notice that by choosing smaller and smaller values for h which *approach zero but never reach it*, the corresponding secant lines approach the thick line in Figure 2.2. We'll call this line the *tangent line* to emphasize that when we zoom in close enough it touches the graph at only one point.

Geometrically, we now have a way to make sense of an *instantaneous* rate of change (IROC): it's simply the slope of the tangent line. I'll denote this slope by $m(a)$, where a is the x-value of the point of tangency (in Figure 2.2, $a = 8$). And, since we arrived at this IROC concept by computing the AROCs of smaller and smaller values of h, we'll express this by saying that

$$m(a) = \lim_{h \to 0} \frac{P(a + h) - P(a)}{h}, \tag{4}$$

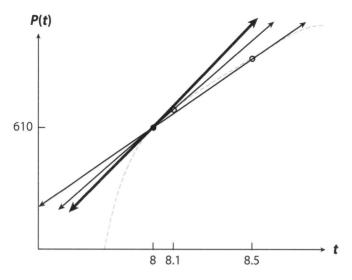

Figure 2.2. AAPL stock price zoomed in around $t = 8$.

where the right-hand side of the equation is read "the limit as h approaches zero of the average rate of change of $P(t)$ between $t = a$ and $t = a + h$."

Mathematicians call the number $m(a)$ *the derivative at $t = a$* of the function $P(t)$. We often suppress its geometric significance and denote it by $P'(a)$ (read "P prime of a") instead of $m(a)$. I chose to keep the $m(a)$ notation to remind us of how we got here: we calculated the slopes of secant *lines* as the intervals got smaller and smaller, and so it should be no surprise that what we get is also a line (the tangent line), and thus has a slope (the IROC). The derivative $P'(a)$ describes how Apple's stock price changes from instant to instant. But by changing the function in formula (4) we can similarly describe the instantaneous change of just about anything. The derivative, therefore, is the "change function" we've been looking for.

Coffee Has Its Limits

The fact that the derivative represents an instantaneous rate of change makes it a widely applicable concept. I'm reminded of this as I step into my kitchen and start thinking about breakfast. The dominant function in this room is *temperature*, $T(t)$. But despite having *nothing* to do with AAPL's stock price, the beauty of mathematics is that we can understand changes in both by using their average and instantaneous rates of change.

If you're like me, the moment you get into the kitchen you enter multitask mode. Most mornings I fire up the stove and start making eggs or oatmeal. Meanwhile, I put together a sandwich for lunch and toast it in the oven. And of course, while all of this is going on the kitchen is slowly being filled with the aromatic smell of brewing coffee. All of this "change" foreshadows the presence of derivatives. And because it smells so good let me focus on the coffee for now.

I'm not much of a coffee drinker, but since an estimated 50% of Americans drink coffee,[12] it's not surprising to me that Zoraida is. What does surprise me is just how quickly coffee gets cold. Pour it in a cup and about 10 minutes later it's cooled down to room temperature. So, for my first act, let me tell you about the derivatives hidden in your morning cup-o-joe.

Let's measure the temperature T of the coffee in degrees Fahrenheit and time t in minutes since the coffee pot left the plate warmer of the coffee maker. My coffee maker keeps the coffee at about 160°, so that $T(0) = 160$. Pretend that I have excellent dexterity and a nanosecond after removing the pot I've poured a cup of coffee, without spilling a drop. After 2 minutes I put a thermometer in the coffee and it measures a temperature of 120°, making $T(2) = 120$. The last bit of information we'll need is that the air in my kitchen is at a temperature of 75°.

Using all of this data, a formula for the temperature $T(t)$ is[ix]

$$T(t) = 75 + 85e^{-0.318t}. \tag{5}$$

The graph of this exponential function is shown in Figure 2.3 for $0 \leq t \leq 25$.

The first thing we notice is that the temperature drops very rapidly within the first 10 minutes, and then drops much more slowly

---

[ix]This formula originates from Newton's Law of Cooling, which belongs to the subject of differential equations.

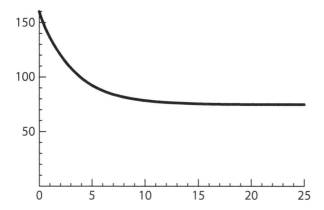

Figure 2.3. The temperature $T(t)$ of the coffee since leaving the warming plate.

afterwards, seeming to eventually reach 75° (which should be no surprise since this is the ambient temperature of the room). But now that we've talked about derivatives, let's try to describe the change in $T(t)$ from instant to instant.

For starters, we can visually see that the tangent lines for $0 \leq t \leq 5$ have pretty steep negative slopes. This tells us that $T'(t)$—the IROC t minutes after removing the coffee from the plate—is negative, reflecting the fact that the coffee's temperature is *decreasing*. This is useful information, but I want to know what I'm up against here. So let's do better than just qualitative observations; let's *compute* how fast the temperature was dropping at the instant I took the pot off of the warming plate.

Mathematically, we're after $T'(0)$; to find it we'll use our formula (4), with $a = 0$ and replacing the function $P(t)$ with $T(t)$. After some simplification,[*3] we're left needing to calculate the limit

$$\lim_{h \to 0} \frac{85(e^{-0.318h} - 1)}{h}. \tag{6}$$

Fear not, for we know (or could go back and reread) how we came up with this notation to begin with: we imagined ourselves *calculating* lots of average rates of change with smaller and smaller h values. That's precisely what we'll do here to calculate the limit in expression (6).

TABLE 2.1.
The limit table for $\lim\limits_{h\to 0}\frac{85(e^{-0.318h}-1)}{h}$.

h	$\frac{85(e^{-0.318h}-1)}{h}$
0.1	-26.6047
0.01	-26.9871
0.001	-27.0257
0.0001	-27.0296
0	undefined
-0.0001	-27.0304
-0.001	-27.0343
-0.01	-27.073
-0.1	-27.4644

First plug a nonzero value for h into the fraction and record the result. Then do this again with a *smaller*, but still nonzero, value for h. This process results in the *limit table* shown in Table 2.1. The hope is that as we plug in smaller h values the numbers we get approach *one* number. And, as the numbers in Table 2.1 hint, as h approaches zero the AROCs are approaching the number -27.03. Congratulations! You've now learned the first method we mathematicians use to calculate limits.

Well, technically all we've done is to estimate the limit, since we don't know that for *even smaller* values of h the AROCs don't approach some *other* number. This observation encourages us to try to find another, less error-prone method of calculating limits (which we'll do shortly). For now, trust me when I say that $T'(0)$ is indeed approximately -27.03.

At this point, I need to interject and make a public service announcement. One might be tempted to conclude that, since the IROC is $-27.03°$ per minute, it follows that one minute later the coffee's temperature has fallen by $-27.03°$. This sure seems reasonable, *but it's not true*: you can check it by computing $T(0) - 27.03$ and comparing it to $T(1)$ from formula (5). The simplest reason why is that the rate at which the temperature is falling is *not* constant over that one-minute interval, as our secant line slope analysis of Figure 2.3 showed.

In fact, we could go back and rework our limit table to calculate $T'(0.1)$, $T'(0.4)$, and every other IROC for every value of t. If we assembled the results into a graph we'd obtain Figure 2.4.

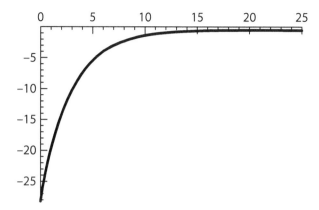

Figure 2.4. The derivative *function* $T'(t)$ of the coffee temperature $T(t)$.

The y-values of the graph in Figure 2.4 tell us the derivative of $T(t)$ at that given t-value. Right away we recognize the y-value of -27.03 at $t = 0$; it's $T'(0)$. We also see that as t increases the graph gets less negative. This agrees with what we found by analyzing the slopes in Figure 2.3. So what we've created in Figure 2.4 is the derivative *function* $T'(t)$.

This function, being literally the collection of the numbers $T'(t)$ for each t between $t = 0$ and $t = 25$, tells us how the coffee's temperature changes at each instant of time. But it does so in a different way than what we've done thus far: in Figure 2.3 we extracted information about $T'(t)$ by calculating slopes of tangent lines, but in Figure 2.4 the y-values are themselves those tangent line slopes.

You might be wondering how I obtained the graph in Figure 2.4. "Surely you didn't calculate thousands of secant line slopes and glue them together" you might say, and you're right, I didn't. There's a much faster way, but we'll have to wait until the next chapter to discover it. Also, since it's clear from Figure 2.4 that $T'(t)$ changes as t changes, you might also say "wait a minute; you said that the derivative, $T'(t)$ in this case, *was* the change function we were looking for; so what function describes the 'change of the change'?" Good question; the answer will occupy a good portion of the next chapter, but let me finish my breakfast (and reheat my now cold coffee) first.

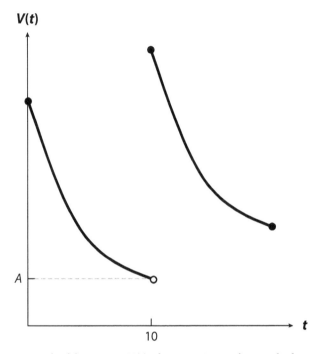

Figure 2.5. A graph of the amount $V(t)$ of vitamins/minerals in my body as a function of time t.

A Multivitamin a Day Keeps the Doctor Away

Like many people I take multivitamins every day, and I take the first one right now, with my breakfast. These little wonder pills—at least the ones I've settled on now—are basically powdered forms of whole foods that release the vitamins and minerals of the produce used to make them into your bloodstream as the pill is digested. What's mathematically noteworthy is that the amount V of total vitamins/minerals in my body as a function of time t (measured in hours) has a graph (Figure 2.5) that looks different from all of the ones we've discussed thus far.

If we let $t = 0$ be the moment that I took my morning vitamin, then as my body breaks down the pill throughout the day, the amount of nutrients left diminishes. Therefore, this amount is changing with time t. So we could certainly think of the instantaneous rate of change of the level of nutrients in terms of derivatives. But when I take another vitamin at dinnertime, roughly 10 hours later, something interesting happens. At that point, I instantaneously bring the total available nutrients from the vitamins I've taken back up.

This is depicted in Figure 2.5 as the "jump" at $t = 10$; we call this a *discontinuity* in the graph, and say that $V(t)$ is *discontinuous* at $t = 10$. The terminology here comes from contrasting the graph of Figure 2.5 with the graphs in Figures 2.1 through 2.4; all of those graphs are *continuous*.

Now, even though at $t = 10$ the function $V(t)$ has changed values, the derivative $V'(10)$ does not exist. We can see this from our definition:

$$V'(10) = \lim_{h \to 0} \frac{V(10 + h) - V(10)}{h}.$$

For positive h, the limit table consists of the slopes between the point $V(10)$ and points on the graph in Figure 2.5 *to the right of 10*. But these lines slant downward, and so are all negative. However, for negative h the limit table will consist of slopes between $V(10)$ and points *to the left of 10*. These are all positive (and large) slopes. Therefore, no matter how small h is, these two sets of numbers will never approach *the same number*.

This analysis illustrates that derivatives are not defined at points of discontinuities. So how do we know when we're dealing with a discontinuous function?

A preliminary answer comes from comparing the graph in Figure 2.5 with those of Figures 2.1 through 2.4. What sets these two groups of graphs apart is whether we can draw the graphs in each group without lifting our pencil. But this "drawing definition" of continuity isn't very mathematical. So let's formulate a better definition.

Suppose I wanted to figure out how much of the first vitamin I took was floating around my bloodstream right *before* I took the second one. Intuitively we know that the answer is the y-value of the lowest point of Figure 2.5, which is A. And since we're looking for the value of $V(t)$

right before $t = 10$, what we want is

$$\lim_{t \to 10^-} V(t). \tag{7}$$

To calculate this limit we can construct a limit table for $V(t)$ to determine the limit, just like in the coffee problem. However, since we're only interested in the remaining vitamin amount *before* $t = 10$, our limit table would only include values such as $t = 9.9, 9.99, 9.999$. This explains the "$-$" superscript above the 10 in expression (7); it's there to remind us of this "limit from the left" concept.

The value of the limit we'd obtain from our limit table can be understood visually as the y-value we'd obtain if we crawled along the graph of $V(t)$ from $t = 0$ to *just before* $t = 10$. The result, the y-value A, confirms our intuition.

Limits like (7) are called *one-sided* limits, and as you may have guessed, we could just as easily define the concept of a "right-hand limit." For these limits we'd approach the t-value from values *greater* than it and investigate the y-values we'd get. This limit would have a "$+$" superscript instead of the "$-$" superscript in expression (7), and an example is the limit

$$\lim_{t \to 10^+} V(t). \tag{8}$$

Take a look at Figure 2.5 once more and see if you can figure out this right-hand limit. If you think the answer is the y-value $V(10)$, then you're correct!

With these new concepts, we can now focus our attention on the simple fact that at $t = 10$ the left and right limits are *not* equal. Why is this important? Well, recall that $V(t)$ is discontinuous at $t = 10$. And remember that we also called a function continuous if you can draw its graph without lifting your pen. So if you're now thinking that limits have something to do with continuity, then you're on the right track. In a way this shouldn't be surprising based on our "drawing definition" of continuity, since the presumption there was that there were no jumps that would require you to lift your pen. But this discussion has now provided us with a way to define continuity *mathematically*.

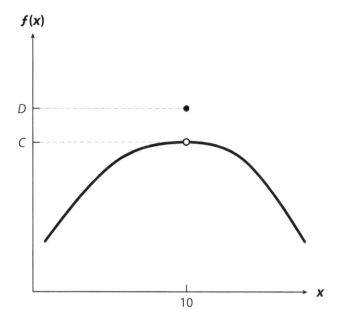

Figure 2.6. A graph depicting a *gap discontinuity*.

Since we want to avoid jumps, we want the left and right limits to be equal. But simply requiring this won't avoid the discontinuity depicted in Figure 2.6.

In this example, the left and right limits as x approaches 10 are the same, and equal to the y-value C. But the value $f(10) = D$, and so there's a *gap discontinuity* in the graph. To avoid this we should require that the value of the function at every point $x = a$ be equal to the common value of the left and right limits as x approaches a.

Putting together the requirements of no gaps or jumps, what we want is the condition

$$\lim_{x \to a} f(x) = f(a), \tag{9}$$

where we no longer write the "+" and "−" superscripts (since we're assuming that both left- and right-hand limits are equal), and we further assume that $f(a)$ is a number.

We have now arrived at the *precise* definition of continuity. A function that satisfies this condition at every point has no jumps or gaps, and therefore we can draw its graph without lifting our pen, recovering our earlier definition of continuity.

Our vitamin discussion has solidified our understanding of limits. We now know that the limit as t approaches a number b of a function $F(t)$ is simply the y-value we reach by walking along the graph of $F(t)$ toward $t = b$. Equivalently, it's the number we obtain from the limit table analysis similar to what we did with the coffee problem. And if the function satisfies (9) at every point in its domain, then it's continuous and its graph can be drawn without lifting our pen. Now imagine if I hadn't taken my vitamin today; look at everything we'd have missed out on. All joking aside, limits form the foundation for calculus (as formula (4) hints), so we've covered important ground here.

Derivatives Are about Change

With all of this talk about vitamins and their absorption rates I neglected to drink my reheated coffee. I could reheat it *again*, but now we're getting silly; anyway, I've finished my breakfast.

The room has gotten darker now; the sun has receded and the clouds outside are turning grayer by the minute. Experience tells me that this change in the weather is typically followed by rain, so I'll pack an umbrella just in case. And now that I think about it, I'll take a light jacket too. Despite July being a hot month in Boston historically, my office tends to be on the chilly side.

In the middle of all of this internal rambling I have a moment of revelation: *change* is ubiquitous in our daily lives. From the stock charts on CNBC to my morning coffee, from the vitamins to the weather, change almost *defines* our lives. And in this chapter we've learned, courtesy of Dr. Newton, that wherever there is change calculus—and a derivative in particular—is not far behind.

||||||||||||||||||||||||||||||||||||||||||||||||||||||||||||||||||||||||||||

CHAPTER 3

DRIVEN BY DERIVATIVES

I'M NOW READY TO HEAD OUT TO WORK. The last thing to do is to grab a pair of shoes from my shoe "collection"—a line of pairs of shoes on the floor ordered by color from light brown to black. Oftentimes I just pick a random pair, but now that it's started to rain I decide to go for the usually neglected pair of waterproof shoes. I slip them on, pick up the umbrella, and finally leave my house.

I open my door and, unlike my shoe collection, I am instantly met with a world of disorder. It's pouring outside; people without umbrellas are rushing for cover, trying to avoid being splashed by the cars passing by. Unlike the EM waves of Chapter 1, there seems to be no unifying theme to it all. But then I realize that everything in this scene, from the raindrops falling to the cars zooming by, is changing. The best I can do right now is to associate a derivative with each type of change in this scene, according to our "wherever there is change there are derivatives" adage from Chapter 2.

But Galileo didn't stop his investigations when he derived a formula. As we learned in Chapter 1, he didn't simply "mathematize" motion, his formulas helped him to *understand* motion. From his formulas he was able to *deduce* parabolic motion, something no one before him had been able to explain. I often urge my students to follow Galileo's lead and let the mathematics they're studying "speak to them" so that they can sit back, listen, and learn. I think now is a good time to take my own advice; I start looking at all the commotion outside in a new way, now determined to use derivatives to better understand what's really going on.

Why Do We Survive Rainy Days?

Puddles of water are starting to form everywhere, but I slither carefully around them on the way to my car. Meanwhile, thousands of raindrops are splatting all over my umbrella. Yet somehow my umbrella is keeping me safe from these hundreds of drops falling from thousands of feet. This isn't particularly surprising *at first*. But then I start thinking about the journey that just *one* of these drops makes.

A typical raindrop falls from an average height of 13,000 feet. Along its fall it combines with other droplets, much like snowballs do in cartoons as they roll down the side of a mountain. This process, known as *coalescence*, increases the droplet's size and mass. As it falls, the droplet's speed also increases. So, if both the droplet's mass *and* its speed are increasing as it falls, how come it doesn't crash right through my umbrella? And how do you figure that I'm still alive after my umbrella is bombarded by *thousands* of these raindrops? There must be a way that derivatives can help us answer these questions.

For simplicity, let's start by thinking of a droplet as a sphere. Since the droplet fuses with other droplets as it falls, its mass increases, and the more massive the droplet becomes the more likely it is to fuse with other droplets (think "snowball effect"). Aha! The mass of the droplet is *changing*. By our adage in Chapter 2, a derivative must be lurking around here. Let's find it by "mathematizing" the problem.

Let's denote by $m(t)$ the mass of the droplet at time t, which we'll measure in seconds. The increasing mass tells us that $m'(t)$—the instantaneous rate of mass increase—is positive.[*1] Now let's mathematize the coalescence phenomenon. Intuitively, we expect large droplets to fuse with other droplets more often than small droplets. In other words, the rate at which a droplet's mass increases depends on how massive it currently is. The mathematical statement of this is that $m'(t)$ is proportional to the mass $m(t)$ of the droplet:[x]

$$m'(t) = 2.3m(t). \tag{10}$$

---

[x]The number 2.3 comes from experiments.

This formulation passes our $m'(t) > 0$ test, since the mass of our raindrop is positive. What we've yet to mathematize is the effect of the increasing speed of the droplet. But an object with a mass and a velocity has *momentum*.

Momentum is another one of those things that we're all familiar with. The picture that comes to my mind is that of a football player about to catch the kickoff ball. The players on the opposing team, intent on tackling the receiver, are traveling very fast. Their combination of large mass $m(t)$ and high velocity $v(t)$ makes for a large amount of momentum, defined as the mass of an object multiplied by its velocity: $m(t)v(t)$. The great scientist Isaac Newton gave us a mathematical way to link changes in an object's momentum with the forces acting on the object. I'm alluding here to Newton's second law:

$$F_{\text{net}} = p'(t), \quad \text{where} \quad p(t) = m(t)v(t) \quad \text{is the object's momentum.}$$
$$(11)$$

This law of physics states that the net forces F_{net} acting on an object with mass $m(t)$ cause a change $p'(t)$ to that object's momentum $p(t)$.[xi] It applies to all objects, whether they be football players or raindrops. Lucky for me the raindrops aren't as massive as football players. But because they still fall from thousands of feet, shouldn't $v(t)$ be very large at the instant the drop hits my umbrella? After all this theory, we still can't figure out why any of us survives *even one* raindrop.

Let's follow Dr. Newton and investigate the change in the droplet's momentum as it falls. The force of gravity pulls the droplet down toward Earth, and from the $F_{\text{net}} = ma$ form of Newton's second law we can express that force as $F_g = m(t)g = 32m(t)$, where $g = 32$ ft/s^2 is the acceleration of gravity. Equation (11) then gives

$$p'(t) = 32m(t). \tag{12}$$

[xi]The more recognizable form of Newton's second law is $F_{\text{net}} = ma$, where a is the object's acceleration. For an object whose mass is constant, $p(t) = mv(t)$, and so $p'(t) = mv'(t) = ma$, since acceleration is the derivative of velocity. Thus these are two equivalent formulations of Newton's equation for our problem.

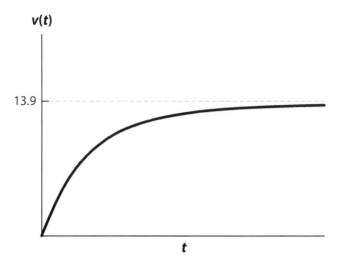

Figure 3.1. A graph of the velocity of a falling raindrop.

Using this equation in combination with equation (10) we can finally determine the velocity function:*2

$$v(t) = \frac{32}{2.3}(1 - e^{-2.3t}). \tag{13}$$

Now that we've done all the math, it's time to let it speak to us. Punch this $v(t)$ into a graphing calculator (or the website wolframalpha.com) and you'll get the graph shown in Figure 3.1. The graph shows that the droplet's velocity is getting closer and closer to $32/2.3 \approx 13.92$ ft/s as it falls. In fact, it seems not to be able to *exceed* that velocity. We can prove this mathematically,*3 but more importantly, this result tells us that eventually something prohibits the droplet from continuing its increase in velocity. What causes this? To answer this question we'll use a technique that Einstein loved: a thought experiment.

Imagine yourself inside a car driving down the freeway, your speed high enough that when you roll the window down you hear the roar of the wind. You put your arm out the window, palm facing down, and don't really feel your arm pushed back. But as you start rotating your palm so that it's eventually perpendicular to the ground, your arm gets flung backwards. The force pushing your arm back is the resistive force of air drag. The strength of that force increased only when you increased the *area* exposed by rotating your palm.

Analogously, as a droplet falls and its size increases, its surface area also increases (recall the snowball analogy). Like your palm, this (larger) surface experiences more air drag, slowing the acceleration of the droplet. Eventually it reaches an acceleration of zero, meaning that its velocity stops increasing. This is what we call *terminal velocity* and is the approximately 13.92 ft/s (or 9.5 mi/hr) shown in Figure 3.1. In a nutshell, air drag is the resistive force equation (13) tells us is present.

But what about the whole coalescence thing? Wasn't the droplet's mass increasing with seemingly nothing to stop it? Here again air resistance saves the day. The air resistance the droplet experiences breaks it up into smaller parts multiple times during its fall. By the time it hits my umbrella, a typical droplet weighs only about 0.0007 pound. In the worst case scenario where the droplets never break up, our analysis shows that they are traveling about 9.5 mi/hr when they hit my umbrella. With such a tiny mass, each raindrop's momentum is just enough to "splat" on my umbrella, but nothing more. Air resistance, therefore, saves us all. Too bad it can't save me from the puddles too, but then again that's where my waterproof boots kick in.

Politics in Derivatives, or Derivatives in Politics?

I made it to my car, safe and sound, but now I'm juggling my umbrella, keys, and bag as I try to get inside the car. Once inside, my trek through the rain is hardly over. Rain is statistically the worst type of weather to drive in, accounting for the vast majority of weather-related crashes.[13] It also reduces average speed by 3% to 13% on freeways. Luckily I've left with a bit of a time cushion so I'm confident I won't be late to work.

I turn my radio on and tune in to old-trusty: WBUR-FM.[xii] The reporter on the air is talking about the unemployment rate. The Democrats are pointing out that recently the rate has begun to decrease from its high, a good sign for the economy. However, the Republicans are countering by stating that the rate of decrease is slowing, a potential problem going forward. That last statement really catches my ear; it isn't a statement about change; it's a statement about how the change is changing. Knowing by now that this should involve derivatives, how can we understand the *change* in a derivative? Is it the derivative of the derivative? Let's get to work by mathematizing these questions.

Let's call the unemployment rate $U(t)$. We know that $U'(t)$ describes the change in $U(t)$, but how do we describe the change in $U'(t)$? Quite simple actually: just call $U'(t)$ something else, like $V(t)$, and reread the second sentence above, replacing U' with V. It now reads: $V'(t)$ describes the change in $V(t)$. Translated back to U's, this says that $(U')'(t)$ describes the change in $U'(t)$. And what is $(U')'(t)$ you ask? It's the *second* derivative, and we usually just write it as $U''(t)$. This new object describes the change in $U'(t)$, just like $U'(t)$ describes the change in $U(t)$.

Okay, with that letter shuffling over, here's a comforting fact: everything we know about the relationship between $U'(t)$ and $U(t)$ is equally valid for the relationship between $U''(t)$ and $U'(t)$. To get back to the unemployment rate, let's note that the recently decreasing rate tells us that $U'(t) < 0$.[*1] But as the Republicans point out, this rate has been slowing. If $U'(t)$ is negative (say -10), but getting less negative (say -9), then its an *increasing* function. Therefore, the change in $U'(t)$, now known as $U''(t)$, is *positive*.

We've seen this type of change before: it's exactly what was happening to the temperature of my coffee in Chapter 2. In effect, the Republicans are saying that the shape of the graph of the unemployment rate $U(t)$ is similar to the graph in Figure 2.3! Who knew that my coffee's temperature could have anything to do with politics and the unemployment rate? You did, of course, since you know that wherever there is change there are derivatives.

[xii]I also listen to music, lest you think me a (total) news junkie.

What the Unemployment Rate Teaches Us about the Curvature of Graphs

What we've just done is *truly remarkable*. By using information about $U''(t)$ and $U'(t)$ we made an educated guess as to the graph of $U(t)$. This is a complete reversal of what we've done thus far, where we have always started with the function and *then* calculated its derivative. So it seems like something deep is going on here; the mathematics is teasing us with new, potentially important relationships between a function and its first and second derivatives.

Let's figure out what's happening by starting with a completely equivalent definition of $f'(a)$:[*4]

$$f'(a) = \lim_{x \to a} \frac{f(x) - f(a)}{x - a}. \tag{14}$$

Now, if x is close to a, then we have[*1]

$$f'(a) \approx \frac{f(x) - f(a)}{x - a}, \quad \text{or} \quad f(x) \approx f(a) + f'(a)(x - a). \tag{15}$$

To better understand this approximation, let's turn to a visual representation of what's going on, depicted by Figure 3.2.

You'll notice in Figure 3.2 that the quantity $f(a) + f'(a)(x - a)$ is of the form $b + m(x - a)$ and therefore a linear function of x. But it's not just any linear function; it's precisely the equation of the tangent line at the point $(a, f(a))$.[*5] Therefore, the approximation in (15) is actually approximating the y-value $f(x)$ (the star in Figure 3.2) by the y-value on the tangent line (the highest dot in Figure 3.2). This process is referred to as *linearization*, and for this reason we say that *the derivative $f'(a)$ linearizes the function $f(x)$ near $x = a$.*

Let me go off on a slight tangent here (a classic calculus pun) and illustrate just how cool linearization is. While I've been talking about the unemployment rate I've been driving to work, which happens to be in a direction away from the radio tower that WBUR-FM uses to broadcast its signal; I estimate that I'm currently about 5 miles away

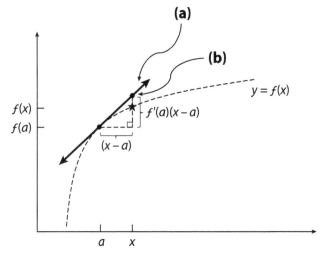

Figure 3.2. (a) This is the line tangent to the graph at the point $(a, f(a))$. (b) The y-value of this tangent line at the point x is $f(a) + f'(a)(x - a)$. We see that this y-value is a close approximation to the actual value $f(x)$ of the function at x (the star in the graph).

from the tower. But in a minute I'll be 6 miles away. From the intensity function $J(r)$ in equation (1) of Chapter 1, I know that this change in distance from the tower will decrease the intensity of the WBUR-FM signal my radio receives, but by how much? Let's use linearization to approximate the change in $J(r)$.

First, we have that $a = 5$, $x = 6$, and $f(x) = J(x)$ (from equation (1) of Chapter 1). Equation (15) then tells me that the change in intensity, $J(x) - J(a)$, is approximately the derivative $J'(a)$ multiplied by the change in distance $x - a$, so that[*6]

$$J(6) - J(5) \approx J'(5)(6 - 5) = -5.9 \times 10^{-6} \quad \text{W/m}^2.$$

The first thing to notice is that this number is negative. This tells me that the intensity is decreasing as I move away from the tower, which makes sense. The second thing to notice is just how small a number it is (5.9 divided by 1 *million*). With such a small change in intensity, and since the station is coming through clearly, this calculation tells me that I can keep enjoying the news throughout my drive.

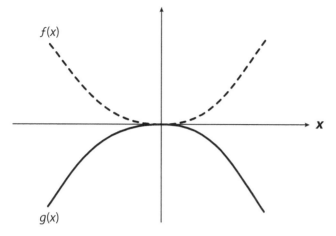

Figure 3.3. The graphs of $f(x) = x^2$ and $g(x) = -x^2$ near $x = 0$.

Thus far we've discovered that $f'(x)$ (1) gives us information about the slope of the graph at the point of tangency and (2) linearizes a function near that point. But we've yet to relate this new knowledge to $f''(x)$. For a clue, let's consider the two functions $f(x) = x^2$ and $g(x) = -x^2$. Their derivatives at $x = 0$ are[*7]

$$f'(0) = 0, \quad f''(0) = 2, \quad g'(0) = 0, \quad g''(0) = -2. \quad (16)$$

Since both have zero derivatives at $x = 0$, linearization tells us that near $x = 0$ both graphs look flat (i.e., have zero slope). But as we can see in Figure 3.3, the graph of $f(x)$ curves upwards, while that of $g(x)$ curves downwards. The first derivative didn't detect this difference; this is a clue that maybe the *second* derivative has something to do with the *curvature* of the graph.

To figure out the relationship, consider the graph in Figure 3.4. Since $f''(x)$ is the derivative of $f'(x)$, then if $f''(x) > 0$, we know that $f'(x)$ is increasing. Since these are the slopes of the tangent lines to the graph of $f(x)$, this tells us that $f(x)$ is curving upwards (Figure 3.4). The reverse situation, when $f''(x) < 0$, similarly tells us that $f(x)$ is curving downwards.

In calculus, we call a function *concave up* whenever $f''(x) > 0$ and *concave down* whenever $f''(x) < 0$. For example, the function in

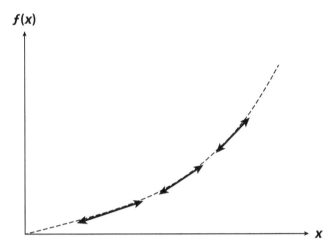

Figure 3.4. Illustrations of how f, f', f'' interrelate. Notice that if $f''(x) > 0$ (as it is for this entire curve), then $f'(x)$ (the slope of the tangent line at x) is increasing. Therefore, as x increases the graph of $f(x)$ has steeper slopes, making it curve upwards.

Figure 3.2 is concave down, and the function in Figure 2.3 is concave up. If at some point $x = c$ the concavity of a function changes, we call the point $x = c$ an *inflection point*. For example, the function in Figure A.3(a) has an inflection point at $x = 0$.

All of this new mathematics was inspired by the discussion of the un-employment rate on my radio. And now that we've "listened" to $f''(x)$ tell its story, what else can we learn from the new mathematics we've developed? Is there a *physical* interpretation of the second derivative (as opposed to the mathematical "curvature of the graph")? Let's get back to my drive to work to find out.

America's Ballooning Population

Thus far my drive has been taking a bit longer due to the rain, but I expected that. Something else I expected that I'm running into now is that dreaded seven-letter word: traffic. In no time I'm bumper-to-bumper moving at a snail's pace.

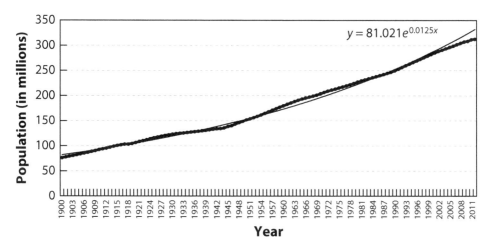

Figure 3.5. The population of the United States since 1900. Retrieved from http://www.ceusus.gov/popest/data/historial.

Nationwide, the average driver in the United States spent 38 hours stuck in traffic in 2011,[14] which translates to about 8.8 minutes for every work day in 2011. Nine minutes might not seem like a lot, but to most of us city folk that feels like an eternity. With the huge financial losses tied to being in traffic—estimated at $121 *billion* in 2007[15]—why haven't we fixed this yet? As I inch forward another foot, counting down from my roughly nine-minute average, I look around and spot the obvious answer: too many cars! Sure, carpooling would help but only to a certain extent. The more fundamental problem is the rising population level. This change has a derivative associated with it, but like the unemployment analysis, can the *second* derivative of the population level give us more information? To find out, let's mathematize the problem.

Let's begin with a graph of the U.S. population since 1900 (Figure 3.5).[16] On the graph I've included an exponential function that roughly fits the curve. Why exponential, you may wonder. There's a simple explanation.

Imagine you're a scientist staring into a microscope. You see a petri dish that contains a single bacteria cell. Bacteria grow and divide quickly, doubling as fast as every 10 minutes. So after 10 minutes your single bacteria cell has doubled to two bacteria. We'll keep track of the two bacteria in the petri dish by writing two as 2^1. When you come back from your coffee break, 10 minutes later, there are now four bacteria in total, or 2^2. Throughout the day, while you work on your other experiments, these four bacteria will themselves double, and the resulting 2^3 will double, and so on. After x 10-minute periods, there'll be 2^x total bacteria cells. In a nutshell, that's why population growth often follows an exponential curve.

   The equation of the curve in Figure 3.5, $y(x) = 81.021e^{0.0125x}$, at first tells us one basic fact: the U.S. population is growing by about 1.25%. But the mathematics contains *much more* information. For example, in the language we've developed the graph of $y(x)$ is increasing and concave up. This means that the first and second derivatives of $y(x)$ are both positive. Therefore, the population level isn't only increasing $(y'(x) > 0)$, it's also increasing at an increasing rate $(y''(x) > 0)$. In particular, this means that more people will be added to the population next year than were added last year. No wonder the roads are getting more crowded!

Feeling Derivatives

After a few more minutes of traffic, my odometer finally reaches a number higher than 5 mph. My velocity $v(t)$ is now increasing, so that $v'(t) > 0$. And since velocity is the derivative of position, $v(t) = s'(t)$, then $v'(t) > 0$ means that $s''(t) > 0$. In other words, I'm now *feeling* the second derivative $s''(t)$. How, you might ask? Let's consult Dr. Newton one more time.

   For fun, picture Newton, complete with his elaborate wig, inside a NASCAR race car. These cars can accelerate from rest to 60 mph in as little as 4 seconds. The one Newton is in is about to take off. Before it does, Isaac feels no forces from the chair he's sitting in. But the instant the car starts moving its velocity changes. Since the car is now accelerating its momentum p is changing, making p' nonzero. By Newton's own second law, this leads to a force F from the seat that pushes against him. From Newton's point of view, he feels pushed back into the seat.

Now, since an acceleration function $a(t)$ describes changes in a velocity function $v(t)$, we know that $a(t) = v'(t)$. And, since we know from before that $v(t) = s'(t)$, putting these two together we arrive at

$$a(t) = s''(t). \tag{17}$$

Thus *acceleration is the second derivative of the position function.* This is the physical interpretation of $f''(x)$ we were looking for; we now see that the second derivative tells us the acceleration of an object with position function $s(t)$. Therefore, we can summarize Newton's thrill ride as follows: every time you feel pushed back against your seat you are *feeling* the second derivative of your car's position function $s(t)$.

The Calculus of Time Travel

Unfortunately for me, more traffic jams loom on the horizon. I don't think I can handle another nine minutes of this, so I start thinking about finding an alternate route to work. For the majority of human existence, we found alternate routes by asking other people for directions; these days I can ask the minicomputer in my car. This gadget relies on the GPS system—an acronym for the Global Positioning Satellite network—of satellites that help the device figure out where I am and how to get me where I'd like to go. One push of a button later, my GPS device locates me and displays a map of my surroundings; it starts tracking my car's location and then pulls up an alternate route. In no time I'm on a less congested road, with my GPS unit continuously guiding me through the unfamiliar new route. Many of us, as I just did, take this device for granted every day. But a spooky story is hidden here, and—believe it or not—it will completely change the way you think about reality.

The first thing to know is that my GPS unit communicates with the GPS satellites in space through signals that travel at the speed of light c, a remarkable 11,176,920 miles *per minute*. These satellites orbit the Earth at about 8,666 miles per hour, *way* slower than the speed of light. Nonetheless, in 1905 Einstein showed that a lot of strange things happen when objects *do* travel at speeds close to *c*.

To put Einstein's findings in perspective, imagine two identical clocks, one on board an airplane traveling with velocity *v* and the other on the ground. Suppose that the pilot measures the flight time to be *y* hours (using her clock). Einstein discovered that back on Earth, the flight time (according to the ground clock) is *not y* hours, but instead *z* hours, where

$$ z = \frac{y}{\sqrt{1 - \frac{v^2}{c^2}}}. \tag{18} $$

Since v^2 is positive, the denominator in this equation is smaller than one. This tells us that *z* is *larger than y*. In other words, Einstein showed that clocks traveling with velocity *v* are slow relative to stationary clocks, a phenomenon now called *time dilation*.[xiii] Put a bit more alarmingly, *moving objects time travel into the future* relative to stationary objects.

To appreciate the bizarre implications of this finding, imagine you take a three-hour trip on a *very fast* plane that can travel at 86% the speed of light. When you return, equation (18) predicts that the world will have aged *six* (not three) hours. This wouldn't be an illusion; every clock and every person would be *six* hours older, even though you're only three hours older (according to your wristwatch). You would, in a very real sense, have traveled three hours into the future,[xiv] and relative to everyone else, you'd be three hours *younger*.

Before you get too excited about this fountain of youth, you should know that speeds close enough to *c* to make this effect noticeable are currently produced by physicists only in particle accelerators— machines specially designed to study the particles that form the

---

[xiii] This is part of Einstein's Special Theory of Relativity. Supposedly, when Einstein was asked to explain relativity in layman's terms, he said "put your hand on a hot stove for a minute and it seems like an hour. Sit with a pretty girl for an hour and it seems like a minute. That's relativity." I can confirm this to be an accurate description of the theory.

[xiv] As this example shows, time travel into the future is completely within the laws of physics. In fact, cosmonaut Sergei Avdeyev holds the world record for time travel into the future, traveling a whole millisecond into the future as a result of his 748 days on board the Mir space station.[17]

building blocks of our universe. To the rest of us nonphysicists, Einstein's discovery might seem irrelevant to our daily lives. Enter my GPS unit.

Let's start by finding out how much time, on Earth, corresponds to one second as measured by a GPS satellite's onboard clock. Since these satellites move at about 0.0013% the speed of light, we can use our results on linearization to rewrite equation (18) for speeds that are slow compared to the speed of light:*[8]

$$z \approx y \left(1 + \frac{v^2}{2c^2} \right). \tag{19}$$

With $y = 1$ and $v/c = 0.000013$, this approximation tells us that 1 second on a GPS satellite's onboard clock is actually about 1.0000000000834623 second to us back on Earth; after one day the discrepancy totals about 0.00000721122 second. This is such a small amount that it seems inconsequential, but remember that the GPS satellites' signals to my GPS unit are traveling at the speed of light, which, as we saw before, is very fast. Consequently, this error in *time* measurement translates to an error in *distance* measurement of 0.00000721122c, or about 1.343 miles *per day*. Imagine using your GPS unit for a cross-country drive and discovering that after just one day it can no longer accurately tell you where you are.[xv] Pretty soon, you'd start to regret having bought such a useless gadget. Fortunately for us the GPS satellites' onboard clocks compensate for these effects (or more precisely, the engineers who design these clocks take into account Einstein's findings), resulting in a GPS network that does prove immensely useful.

Thus far I've focused on my GPS unit, but Einstein's equation (18) applies to *all* moving objects. This is where things get truly mind-bending. For example, if your dog runs through the park for 30 minutes, (18) says he's now traveled into the future *relative to you*. But after you drop him off at home and go to the grocery store, you've traveled into

---

[xv] In fact, since these errors would occur every day, the whole GPS network would become useless within a few weeks.

the future *relative to him*. Meanwhile, *everything* you saw in motion around you during your trip to the store—other people, other cars, etc.—traveled into the future relative to *someone or something* else. Sorting all this out would confuse even soap opera writers!

In this chapter we've "mathematized" several aspects of our daily lives—from rainfall to driving forces to congestion on the roadways—and have learned much more about them by letting the mathematics speak to us. But come on, time dilation takes the cake. It perfectly embodies our "wherever there is change there are derivatives" adage, *and* it takes it a step further by providing us with a radically new way to look at reality. If these first three chapters are already making you rethink the reality you live in, let me find a parking spot and get up to my office to tell you the rest of the story.

CHAPTER 4

CONNECTED BY CALCULUS

IF YOU'RE LIKE me, one of the first things you do when you get to work is check your e-mail. Honestly, I don't know how some jobs ever got done before e-mail. I can't imagine my students writing me *letters* with their questions on the homework, or me making costly international calls to my collaborators to discuss research. With e-mail it's now become much easier—and faster—to communicate with each other. But it's not just ease of communication that this new technology has produced. In this age of Facebook and Twitter we're all connected to each other. This makes me think about Einstein's time dilation discovery. That one concept, the relativity of time, connects *everything*. I start wondering what other phenomena we can connect with mathematics.

E-Mails, Texts, Tweets, Ah!

In the midst of my internal rambling, a new e-mail pops up on my screen. If you're like me, this happens multiple times a day. In fact, in 2010 an estimated 294 *billion* e-mails were sent *every day*,[18] which equates to about 3.4 million e-mails *per second*! Roughly 90% of these e-mails are spam, and despite our spam filters a few always get through. Sifting through our enlarged inbox prevents us from doing the work we're paid to do as employees, and costs us (and the employers we work for) about $22 billion every year in lost productivity.[19] But how do we quantify "productivity"? And is there anything we can learn from doing so that'll help us be more productive? Let's take the approach of the previous chapter and mathematize the problem.

The time we lose to this "instant messaging" issue could've been spent producing another computer, shirt, or car. So let's think of lost productivity as drop in the dollar value of goods produced. If we call $p(x)$ the total value of the goods a company with x employees produces, then the *average productivity* $A(x)$ of the company's workforce is

$$A(x) = \frac{p(x)}{x}. \tag{20}$$

The function $A(x)$ puts a number on how much a company's x employees produce (in dollar values of a particular product) on average. For example, for a company that manufactures and sells $30 radios and has 10 employees we'd have $A(10) = \$30/10 = \$3$, meaning that, on average, each employee contributes $3 worth of production value to each radio.

Naturally, companies would like their workforce's average productivity to be increasing. In this environment the company could *hire* people and *increase* the dollar value of goods produced. By now we know that in order for a function to increase we need its derivative to be positive, so let's get to work.

Let's use calculus to calculate the derivative of $A(x)$. We get[*1]

$$A'(x) = \frac{xp'(x) - p(x)}{x^2}. \tag{21}$$

We see that $A'(x)$ is positive only when the numerator is positive (since the denominator is never negative). This happens when[*2]

$$p'(x) > A(x). \tag{22}$$

Barely a few lines in and the mathematics is already speaking to us. Here's what it's saying.

Since $p'(x)$ is the instantaneous rate of change of the total value of the goods a company with x employees produces, this condition says that if this rate is greater than the average productivity of the company's workforce, then that average productivity will increase. As you can see, this sentence isn't "user-friendly." We can rephrase it in the following much more helpful way:[*3] if $p(x)$ grows faster than a linear function (i.e., $p(x)$ is quadratic, cubic, etc.), then $A(x)$ will increase. This is *much* more useful, since most companies have mounds of data they can sift through to determine the shape of their $p(x)$ curves. If condition (22) isn't satisfied, the company has a variety of approaches it can take to increase productivity.

One natural approach is to reassign employees to the tasks they're most productive at. So in a very real sense you're connected to your coworkers through your employer's $A(x)$ function. If this function's overall value is too low, you may be reassigned to a new task or project team in an effort to increase output (and therefore, most likely, profit). Here, then, is another example of how calculus connects seemingly unrelated aspects of our lives. In my world, I experience this often as I get asked to attend certain meetings based on past work that I've done. In fact, now that I've finished decluttering my inbox and look at my calendar I notice that I have one coming up.

The Calculus of Colds

My first meeting of the day is with colleagues, students, and administrators. With today's rainy day some of them are walking into the room absolutely soaked. We're inside a "smallish" classroom that seats 30, and my close proximity to some of these shivering souls (my building is usually cold) makes me think of my mother. Let me explain.

As children we were all told by our mothers to avoid getting rained on. To this day my mother insists that getting rained on will give you a cold; turns out there's *some* truth to this, but not for the reasons my mother cites. We now know that the common cold spreads through contact with infected individuals. This doesn't depend on whether it's raining, so why am I still worried about the soaking wet people sitting next to me? The reason is that on rainy days more people stay indoors, increasing the likelihood of bumping into someone who *already* has a

cold. I'm not sure if anyone at my meeting has a cold, but if someone
does, how likely is it that I'll catch it before the meeting ends?

We can start by dividing the 20 people at my meeting (including
myself) into two groups: those who are infected—denoted by I—
and those who are susceptible to infection—denoted by S. Both these
numbers may change during the course of the meeting, so both depend
on time; let's incorporate this into our analysis by promoting I and S
into the functions $I(t)$ and $S(t)$, measuring t in hours. The group size
tells us that

$$I(t) + S(t) = 20. \tag{23}$$

But how do we describe the spread of the infection? Well, when the
infection spreads, $I(t)$ is changing, so cue our derivatives! To make
things concrete, suppose five people in the room have a cold. As they
interact with the susceptible population the disease is likely to spread,
and the rate at which individuals get infected, $I'(t)$, will be larger if there
are more interactions. This is leading us to consider the model

$$I'(t) = kI(t)S(t), \tag{24}$$

where $k > 0$ is a constant that describes how fast people get infected
from these interactions, and the product $I(t)S(t)$ is a measure of how
many interactions could result. Using equation (23) we can rewrite
equation (24) as

$$I' = kI(20 - I), \quad \text{or} \quad I' = 20kI - kI^2. \tag{25}$$

This equation is an example of a *logistic equation*.[xvi] We can verify that
the solution to this equation—the number of infected individuals—is[*4]

$$I(t) = \frac{20}{1 + 3e^{-20kt}}. \tag{26}$$

[xvi] A general logistic equation looks like $p' = ap - bp^2$, where a and b are numbers. In 1837 the
Dutch mathematical biologist Pierre-François Verhulst introduced this mathematical model (with
$a, b > 0$) and used it to describe population growth.

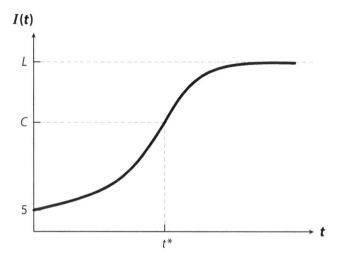

Figure 4.1. The graph of the function $I(t)$ for $k = 1$.

As we can see from the graph in Figure 4.1, $I(t)$ curves upward before t^* and downward after. In the parlance of Chapter 3 this means that $I''(t) > 0$ before t^* and $I''(t) < 0$ after, making the time t^* an inflection point. Moreover, since we know that the second derivative describes the "acceleration" of the function, the change in concavity at $t = t^*$ is telling us something very valuable: the number of infected individuals is increasing at an *accelerating* rate before time t^* and increasing at a *decelerating* rate after that time.

Using my starting assumptions, the number C of infected individuals at time t^* is 10.[*5] So our model is telling us that after half the people in my meeting get infected, the infection rate begins to slow. But what about the value L? What does that tells us?

Well, if by some stroke of (bad) luck I get stuck inside this meeting room for days on end, then we'd expect that *everyone* would eventually catch the cold. This intuition tells us that $L = 20$, a fact that we can verify by using limits.[*6] This is a general feature of the logistic equation when a and b are positive: solutions eventually approach a limiting value L, called the *carrying capacity*.

Luckily, I'm here for only an hour. I've already noticed a few people clearing their throats though, so I'm not in the clear just yet. To get some peace of mind I can calculate the expected number of infected individuals one hour after the meeting starts:

$$I(1) = \frac{20}{1 + 3e^{-20k}}.$$

Notice that this number depends on k, the number that describes how fast the infection spreads. For a relatively normal $k = 0.02$, $I(1) \approx 6.64$, meaning that by the time the meeting ends almost two more people have caught the cold. The problem is that I can't tell who's infected or not. It could be the soaking wet student to my left, or the perfectly normal-looking guy to my right. But either way, our analysis has helped us narrow down the possibilities. This logistic approach to the spread of disease has connected me—quite literally, since I'm one of the $S(t)$—to the other people in the room.

What Does Sustainability Have to Do with Catching a Cold?

The hour is almost up. I've missed most of what was said in the meeting trying to think about this logistic problem and simultaneously avoid whoever looks sick. I don't feel any different as I walk out of the meeting, so maybe I successfully dodged the cold. Between thinking about the logistic equation, trying to identify the possibly infected people, and imagining being stuck in that room for eternity, the one thing I have caught is the hunger bug.

On my way back to my office, I remember that my colleague Stanley told me about a new sushi restaurant that opened nearby. I'm not a fan of raw fish, but I do enjoy the cooked makimono rolls. Plus, it's now turned into a nice day thanks to Nature's usual cycle of rain followed by sunshine. So I decide to go ahead with the sushi plan. I send Stanley a quick text (which probably distracted him, dropping *his* productivity) to see if he wants to join me. A few of minutes later we meet up and start walking over.

When we get to the restaurant I'm struck by just how busy the place is. Servers are rushing by with plates full of seafood, and only a few tables are open. Granted it is a relatively new place, but still the popularity of sushi continues to amaze me. We snag a table and start looking through the menu. I'm struck by how many dishes they offer; there's so much fish in so many combinations to choose from! We order some roll dishes; a few minutes later the server brings us a beautiful arrangement of colors and flavors on trendy-looking plates. While I eat the various rolls, I'm reminded of the population discussion of Chapter 3; there must be *thousands* of other sushi restaurants across the world *each* serving roomfulls of fish *every day*. This presents a huge fishing challenge, and, more important, raises the obvious question: how long can we continue fishing our oceans before no more fish are left to catch? The answer to that question depends, of course, on how much fishing is going on. But it also depends on how many fish there are. So let's mathematize the problem.

Let's call $p(t)$ the world population of fish that humans eat, where t is measured in years. If $p(t)$ is small, then fish are harder to find, making it easy for their population to grow; a simple model here would be exponential growth similar to our bacteria example of Chapter 3: $p'(t) = ap(t)$, where $a > 0$ describes the reproduction rate. But once the population gets too large we (and their predators) slow that population growth. We can account for this by replacing the *constant* growth rate a by the *variable* growth rate $a - bp$ (where $b > 0$), which decreases as the population grows. Therefore, our model becomes

$$p'(t) = p(a - bp) = ap - bp^2. \tag{27}$$

This is a logistic model for $p(t)$! And we've just connected the spread of a cold to the changes in global fish populations! But before we get too excited, we should account for human fishing. If we denote by $100c > 0$ the percentage of the fish population we catch every year, then the modification is simple:

$$p'(t) = ap - bp^2 - cp = (a - c)p - bp^2. \tag{28}$$

In other words, fishing simply lowers the reproduction rate.

The solution to equation (28) is

$$p(t) = \frac{(a-c)p_0}{bp_0 + ((a-c) - bp_0)e^{-(a-c)t}}, \qquad (29)$$

where p_0 is the initial fish population. Now, one of the great things about mathematics is that its conclusions are universal. This means that all of the mathematical conclusions we obtained when we discussed the spread of a cold are valid here as well. For example, let's assume that we don't catch fish faster than they can reproduce, so that $c < a$. Then (29) says that the fish population eventually reaches the limiting value $(a-c)/b$, found again by taking limits;[*7] recall that we called this value the carrying capacity in the cold problem. Moreover, the population of fish will be increasing at an accelerating rate until it reaches half of this value—in this case $(a-c)/2b$—and then continue increasing but at a decelerating rate.

The fact that the carrying capacity is $(a-c)/b$ tells us that the more we fish—the greater c is—the smaller the eventual population of fish will be. But we already knew that. Have we done all this work for nothing? Has my mantra of mathematization failed to provide us with new insights? No way.

Suppose that we now insist that eventually at least M fish survive. From $(a-c)/b > M$ it follows that $c < a - (bM)$. A fishing rate less than this would guarantee, under this model, that enough fish would reproduce to eventually yield a population of at least M fish. This usage of the logistic equation is at the heart of *sustainability analysis*, where the more general question is to determine an approach to harvesting something (be it fish, plants, or even oil) that's sustainable over the long term. And what we've just discovered is that, despite how different these resources are, harvesting them in a sustainable manner can be studied with just one equation: the logistic equation. How cool is that!

What Does Your Retirement Income Have to Do with Traffic?

All of this doom and gloom has left me feeling guilty about eating the (admittedly yummy) food we've just had. And now that my mind isn't clouded by hunger, I leave the restaurant intrigued by how many seemingly unrelated phenomena calculus has thus far managed to connect.

Back at the office I pull up my Internet browser and check my schedule. Off in the corner I catch a glimpse of the financial markets; the Dow is down 1.2%, with similar drops for the other indices. Since the markets go up and down, I've learned not to obsess over these numbers, but sometimes this is easier said than done.

One such time was in 2008, when the market sometimes seemed in free fall. The large drops in the market led many of us to sell everything, no doubt encouraging even more panic selling. With the Federal Reserve calculating that the median net worth of families dropped 39% from 2007 to 2010,[20] many of us walked away from that time period afraid to invest in the markets again. While this may be a good idea for those nearing retirement, if you have decades left before you retire, you might want to think twice about abandoning the market altogether. Allow me to let the mathematics speak for itself.

Let's start by letting $B(t)$ be the balance of your retirement account at the beginning of year t, and let's assume that when your investments make money this gain is immediately reinvested.[xvii] This assumption means that the more money in your account, the more is available to be invested. If your rate of return is $r\%$ per year, then the rate $B'(t)$ at which your balance is changing is

$$B'(t) = \frac{r}{100} B(t). \tag{30}$$

The solution to this equation is $B(t) = B(0)e^{rt/100}$, where $B(0)$ is the starting balance of your account. The math is telling us that your account balance will grow exponentially, the same type of growth we encountered when discussing population growth in Chapter 3. This makes sense, since in Chapter 3 we reasoned that more people around would result in more births and hence a larger population, thus producing even more people and so on. This too is the situation here,

---

[xvii]The technical assumption is that your gains are "compounded continuously."

except that now with more money in your account your gains will be larger, resulting in even more money, and so on. Thus, we've already connected population growth to your retirement account, but let's go even further.

If in addition to your gains you deposit an additional s dollars per year[xviii] then equation (30) becomes

$$B'(t) = \frac{r}{100} B(t) + s. \tag{31}$$

The solution to this equation is[*8]

$$B(t) = \left(B(0) + \frac{100s}{r} \right) e^{rt/100} - \frac{100s}{r}.$$

To see this equation in action, suppose that you are 20 years away from retirement and that you currently have $B(0) = \$30,000$ saved and are adding $s = \$5,000$ per year to your account. Assume further, that $r = 7.2\%$ (we'll see where this comes from in a minute). Then the balance of your account at retirement would be

$$B(20) = \$350, 280.31.$$

But how much of this is due to your yearly $5,000 contributions? Subtracting your total deposits over the 20-year period, as well as the initial account balance, shows that *over 68%* of the $320,280.31 gain over the 20-year period came from the compounding effect (the "gains of the gains" effect).[*9] An even longer time horizon—or rate of return—would only increase this percentage. And why is a 7.2% return reasonable? According to a recent study by Oppenheimer Funds, stock market returns averaged 7.2% over *every* 20-year period between 1950 and 2010 (where the periods begin every month), with all such returns positive.[21] So, for investors with a long time horizon the market's rate of return is quite favorable, even in spite of drops in

---

[xviii]The technical assumption here is that your deposits are made continuously—perhaps as a good approximation every day—and add up to exactly s dollars per year.

the market like that of 2008. Today's 1.2% drop on the Dow isn't anywhere near the drops of 2008, but this analysis does remind me that my time is better spent going about my day than worrying about the market.

The Calculus of the Sweet Tooth

After putting in a good couple of hours of work after lunch, the fact that it's Friday is now setting in. I'm finding it hard to motivate myself to work through the last two hours of the day; I think this is a good time to take a break and head to the coffee shop.

I'm always surprised by how many people bring their work with them to a coffee shop. I usually go to coffee shops to grab some coffee (or a sweet) and *escape* work, not bring it with me. But I guess with the average coffee shop now offering free WiFi, a variety of drinks with seemingly never-ending names, and carefully thought-out, comfortable seating arrangements, I can see how these places are becoming a popular place to get work done. Here, then, is one way a lot of us in the coffee shop are connected: we're here either doing or soon getting back to work. But where's the math?

While I'm in line I put my calculus hat on (figuratively speaking, of course). I see cashiers and baristas, so there's likely to be some productivity-based shuffling of tasks based on the shop's $A(x)$ function; those who make drinks faster likely staff the machines, while other employees who are better at taking drink orders are likely up front. Since I'm in need of a sugar boost I order a hot chocolate, and here's where some literally life-saving math can be found.

These days the barista's machine does all of the work. She puts a scoop of chocolate into the cup, pushes a button, and the machine heats and froths the milk first and then pours it into the cup. The liquid level rises quickly as the machine pours the milk. I know that the flow rate of the milk is constant, but then I notice the liquid level rising slower as more milk is poured into the cup. As we now know, phrases like "changes less quickly" indicate the presence of derivatives. So, if I knew how fast the milk was being poured into the cup, could I determine how

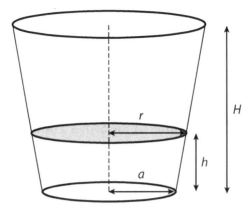

Figure 4.2. The frustum shape of my cup.

the volume of liquid in the cup was changing? As I sometimes tell my students: *por supuesto* (Spanish for "of course")!

Let's start with the shape of the cup; it's a frustum—a cone that's had its tip cut off perpendicular to its height (Figure 4.2). Suppose its smaller radius is a and its larger radius is b, and that it has height H. As the milk enters the cup the volume of milk plus chocolate also forms a frustum. Let's call *its* height h and denote its larger radius by r. The volume of a frustum with radii a and r $(a < r)$ and height h is

$$V = \frac{\pi h}{3} \left(r^2 + a^2 + ar \right). \tag{32}$$

If we imagine taking a derivative of the entire equation, we'll have found a relationship between the rate at which the volume is changing, $V'(t)$, and the rates at which the radius r is changing, $r'(t)$, and height h is changing, $h'(t)$. This problem is a classic example of what we mathematicians call a "related rates" problem, since the rates here are related by the frustum volume equation.

To proceed we need the derivative of the volume equation. But this equation involves two variables (r and h). Mathematically that's okay, but it'd be nice if we could eliminate one variable. If you stare at Figure 4.2 and think back to your geometry class, you might spot something: two similar triangles. We can use these to rewrite the volume equation as[*10]

$$V = \frac{\pi}{3}\left(3a^2 h + \frac{3a(b-a)}{H}h^2 + \frac{(b-a)^2}{H^2}h^3\right). \qquad (33)$$

This single-variable function $V(h(t))$ is now straightforward to differentiate using something called the *chain rule:*[*11]

$$V'(t) = \frac{\pi}{3}\left(3a^2 + \frac{6a(b-a)}{H}h(t) + \frac{3(b-a)^2}{H^2}(h(t))^2\right)h'(t). \quad (34)$$

This may be the scariest-looking equation we've seen yet. But have no fear, the professor is here! And my suggestion? Listen to the mathematics.

For starters, remember that a, b, H are all numbers (they are the radii of the cup and its height, respectively). With this in mind, our equation says that $V'(t)$ is just a quadratic function of h multiplied by $h'(t)$. This tells us that at time t the rate at which the liquid is rising, $h'(t)$, is related to both the volume rate $V'(t)$ at which the machine is pouring in the milk and the liquid level $h(t)$. If we assume that the machine pours in the milk at the constant rate $V'(t) = C$, then by solving for $h'(t)$ we see that

$$h'(t) = \frac{C}{\pi\left(a^2 + \frac{2a(b-a)}{H}h(t) + \frac{(b-a)^2}{H^2}(h(t))^2\right)}. \qquad (35)$$

However ugly this formula looks, it confirms our earlier observation: as the liquid level rises (as $h(t)$ gets larger), the rate at which it rises, $h'(t)$, slows down (since the denominator gets larger, and hence the fraction gets smaller). But this formula does much more than just confirm our observation. This formula tells us *exactly* how fast the liquid level is rising at a certain time t and level $h(t)$. That's great, but it's not life-saving mathematics (as I earlier claimed could be found at a coffee shop). However, like the $r(V)$ nemesis of Edison in Chapter 1, maybe all we need to do is look at our $h'(t)$ formula differently.

With this in mind, I leave the coffee shop, hot choco in hand, and head back to the office. Today's rainy morning is long gone from my memory, but small puddles of water are still left over. They look like miniaturized lakes; and that's when it clicks: the same mathematics we used to find $h'(t)$ given $V'(t)$ for my hot choco applies to these puddles. And since these puddles are miniaturized lakes, we can use the same mathematics to analyze everything from reservoir levels after rainfalls to flooding concerns. For example, emergency management agencies determining whether to order an evacuation due to flooding face a problem similar to my hot choco example. They can measure the rate $V'(t)$ of the volume of rain falling on the area of interest but then need to determine how fast the water level is rising, that is, $h'(t)$. For situations where $V'(t)$ can be estimated in advance—for example, in the case of hurricanes—an equation similar to $h'(t)$ could help determine whether to order an evacuation. The mathematics of finding V might be more complicated, but we've now learned that we can connect the important problem of evacuating a flooding zone to the seemingly silly "hot chocolate problem." This is yet another example of how mathematics makes unexpected connections that, in some cases, may even save lives.

Throughout this chapter we've seen how mathematics, and specifically calculus, connects many different phenomena through its formulas, concepts, and reasoning. Who would've known that the same equation that describes how infections spread also has implications for sustainable fishing? And how cool is it that the mathematics of population growth is also behind the growth of our 401(k)s? Or how about making life-saving decisions with the same mathematics we see every day at the coffee shop? If this were all we could do with mathematics, we'd probably be happy with just that. But I'll interject right here and give my sales pitch: "But wait, there's more!" In the next chapter we'll see how calculus helps us to make life *better*. Like the old BASF commercial, we'll see that calculus doesn't just describe your world, calculus makes your life . . . better.

CHAPTER 5

||||||||||||||||||||||||||||||||||||||||||||||||||||||||||||||||||||||||||||||||||||||||||||||||||||||||||||||||||||||||||||||||||||||||||||||||||||||||||||

TAKE A DERIVATIVE AND YOU'LL FEEL BETTER

MY OFFICE IS ON THE THIRD FLOOR of the building I work in. As I walk in, hot chocolate about half-way gone, I head toward the stairwell. I climb these three flights of stairs many times during the day. Naturally, the first couple of steps are easy, but as I keep moving up the stairs my heart beats faster. It's compensating for the sudden increase in oxygen demand, and quickly distributing that oxygen across my blood vessels to my muscles. But this requires very special plumbing. For starters, my blood vessels need to expand to accommodate the greater volume of blood flowing (if we want to keep the blood pressure down). How much should they expand by? In addition, that blood needs to make it to my muscles as fast as possible. With blood vessels branching in all sorts of directions, this raises another question: how does your body know what the *most efficient* branching directions are? Like our flood-zone application in the last chapter, these are literally life and death questions. Let's discuss the first and come back to the branching issue.

I "Heart" Differentials

In 1838, the French physiologist Jean Louis Marie Poiseuille studied the more general problem of a liquid flowing down a (cylindrical) pipe. He discovered that at any instant in time t the volume flow rate $V'(t)$ of liquid flowing was related to the radius r of the pipe by

$$V'(t) = k(r(t))^4, \tag{36}$$

where the constant k depends on several physically relevant parameters—among them the fluid viscosity.[xix]

But our vessel-expansion problem asks something different: if the volume flow rate *changes*, what's the resulting change in the blood vessel radius r? Notice that this question is about how changes in V' affect r and has nothing to do with time t. So let's pretend that we take a snapshot of one of my arteries at time $t = t_0$ and let's rewrite Poiseuille's equation as

$$f(r) = kr^4, \tag{37}$$

where f is our volume flow rate V', but considered—at the instant of time t_0—as a function of the artery radius r. This new relationship now relates the volume flow rate to the radius, exactly what we want. Further, let's say that the current radius of the artery is $r = a$. In Chapter 3 we discussed how, for values of r close to a, we could approximate the value $f(r)$ by

$$f(r) \approx f(a) + f'(a)(r - a). \tag{38}$$

If we introduce the notation $\Delta f = f(r) - f(a)$ and $\Delta r = r - a$ (read "the change in f" and "the change in r," respectively), then we can rewrite this as

$$\Delta f \approx f'(a)\Delta r. \tag{39}$$

Now, in deriving this approximation we assumed that the change Δr was small (this is equivalent to assuming that r is close to a), but what if you imagine making Δr, and hence Δf, as small as possible but still nonzero (i.e., "infinitesimally small")? What you get is

$$df = f'(a)\,dr. \tag{40}$$

[xix]The viscosity of a fluid is a measure of how much a fluid resists flow. For example, honey has a higher viscosity than water.

The two new objects here, df and dr, are called *differentials*. In calculus-speak, this equation tells us that the infinitesimally small change dr in r causes the infinitesimally small change $f'(a)\,dr$ in $f(r)$. Getting back to Poiseuille's formula for f, we can differentiate to get[*1]

$$df = 4ka^3\,dr. \tag{41}$$

If we now divide by $f(a)$, then

$$\frac{df}{f} = \frac{4ka^3}{ka^4}\,dr = \frac{4}{a}\,dr = 4\frac{dr}{a}. \tag{42}$$

Notice that the quantity dr/a is the change in r divided by its starting value. In other words, dr/a is just the percentage change in the initial artery radius a. Similarly, df/f is the resulting percentage change in f. Thus, our result says us that a 4% increase in the blood flow rate f (a df/f value of 0.04) would result in a 1% increase in the radius of the artery. In fact, as we can see from the equation, the percentage increase in the artery radius r will always be one-fourth the percentage increase in the blood flow rate f.

Now that we've answered our first question, we are beginning to see just how efficient our bodies are. But there's *much* more efficiency built-in, as we'll soon see when we discuss the branching problem. But first I need to tell you about how we mathematize questions like "What is the most efficient branching angle?"

How Life (and Nature) Uses Calculus

On my way down the hall to my office I take another sip of my hot chocolate drink. What's left of it is now cold, having suffered the same fate my coffee did in Chapter 2. Since I'm not a fan of cold hot chocolate, once I walk into my office I throw the cup halfway across the room, hoping to make it inside the trash can. I want you to picture that cup flying across the room in slow motion, as if it were one of those "bullet-time" scenes in *The Matrix*. From our Chapter 1 work and Galileo's genius, we know that its trajectory is a parabola. We also know from experience that what goes up must come down. And this seemingly insignificant fact is actually the doorway into our study of *optimization*, the subfield of mathematics dedicated to maximizing (or minimizing) functions. Let me tell you how.

Pretend that I missed the trash can by a mile and instead threw the coffee cup straight up in the air (maybe I slipped on banana peel). The cup goes up, then it comes down. Okay, fine. But what happens in between? At *some* point it has to switch direction from "going up" to "going down." In other words, at some point it has to be *at rest* (going neither up nor down). This is a pretty radical idea; one would think something thrown in the air is always in motion, but that's not true. What can we learn from this? Let's mathematize and find out.

Let's label the vertical position of the cup at time t by $y(t)$. Then its vertical velocity is just $v(t) = y'(t)$. So, if at some point in time—let's call it t_0—the cup is at rest, then $v(t_0) = 0$. But this is equivalent to $y'(t_0) = 0$. Now, if we look at the trajectory of the cup—depicted in Figure 5.1—we see that $y'(t_0) = 0$ is also the condition for maximum height, that is, for $y(t)$ to be a maximum. This analysis is hinting at something deeper.

If we take the graph of a function $f(x)$ and pretend that it's the graph of my cup's distance to the floor once I throw it in the air, this analysis suggests that to find the maximum value of $f(x)$ we should find where its derivative $f'(x)$ is zero. Let's call these x-values the *stationary points* to keep the velocity analogy intact. The question then becomes: do the maxima (and minima) of a function *always* occur at stationary points?

To answer this question, consider the function $f(x) = x$ for values of x between zero and two: $0 \leq x \leq 2$. Its largest value on this interval is $f(2) = 2$, yet $f'(x) = 1$, showing that f has no stationary points (since $f'(x)$ is never zero). This example teaches us that the maxima and minima of a function don't always occur at stationary points. Moreover, it teaches us something more: the endpoints of your interval matter. For example, if we change the interval to $0 \leq x \leq 3$, then the maximum changes: it's now $f(3) = 3$. So now we have *two* types of candidate x-values for the locations of the extrema: the stationary points (where $f'(x) = 0$), and the endpoints a and b of the interval $a \leq x \leq b$.

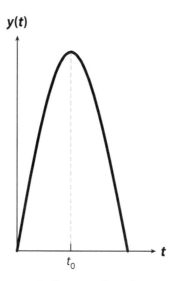

Figure 5.1. A cup thrown vertically upward reaches its maximum height when $y'(t_0) = 0$.

It turns out that if $f(x)$ is a differentiable function, meaning $f'(x)$ exists at each point in the interval $a \leq x \leq b$, then the extrema *always* occur at either the stationary points or the endpoints.[*2]

Our careful analysis of the trajectory of my cup has yielded a strategy for finding the extrema of a differentiable function $f(x)$ on the closed interval $a \leq x \leq b$. First find the stationary points. Then compare the y-values at these points with the y-values of the endpoints a and b; whichever is the largest will be the maximum and whichever is the smallest will be the minimum.

Great, but what does this have to do with Life and Nature? Let's get back to my blood vessels and the branching question. Mr. Poiseuille—and our differentials work—helped us understand why only small dilations in our vessels are needed to accommodate a larger volume of blood flow. But Life is *much* smarter than that. Our arteries don't just want to "accommodate" greater volumes of blood flow; they'd like to *minimize* the work needed to expand those blood vessels. Aha! This is starting to sound like an optimization problem! But we need a function—and an interval—before we start optimizing.

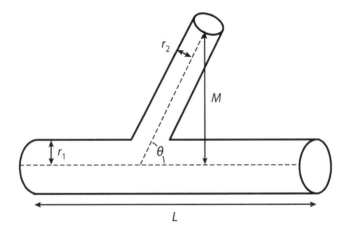

Figure 5.2. A larger blood vessel of length L and radius r_1 branching into a smaller one with radius r_2 at an angle θ.

In his other investigations, Poiseuille came up with a formula relating the resistance R of a liquid traveling through a pipe to its length l and radius r:

$$R = c\frac{l}{r^4}, \tag{43}$$

where c is a parameter that depends on, among other things, the viscosity of the liquid. If our bodies want to minimize the work required to pump blood, our vessels should be configured to minimize the resistance R that blood encounters as it flows. In particular, when blood vessels split into different branches (see Figure 5.2) this branching should minimize R. From this perspective, the question is: what is the optimal angle at which this branching should occur?[22]

By using Poiseuille's second law, we can determine the total resistance blood flowing from the larger vessel up into the smaller would experience:[*3]

$$R(\theta) = c\left(\frac{L - M\cot\theta}{r_1^4} + \frac{M\csc\theta}{r_2^4}\right). \tag{44}$$

Now we need the interval. Figure 5.2 suggests we focus on the interval $0 \le \theta \le \pi$ (where θ is measured in radians),[xx] since angles

---

[xx]An angle that measures π "radians" is equivalent to an angle measure of $180°$.

R(θ)

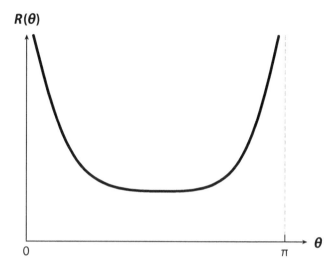

0   π

Figure 5.3. A representative graph of $R(\theta)$.

bigger than $180°$ would correspond to flipping our diagram upside
down and left to right. However, mathematically the endpoints 0 and π
of this interval present a problem, since the functions $\cot\theta$ and $\csc\theta$ are
not defined at those θ-values. But a quick glance at the graph of $R(\theta)$
(Figure 5.3) shows that the minimum of $R(\theta)$ isn't at the endpoints
anyway, since the function shoots off to infinity there. Moreover, since
Figure 5.3 also shows that $R'(\theta)$ exists everywhere inside $0 < \theta < \pi$,
making R a differentiable function, Fermat's Theorem[*2] says that the
minimum value must occur at a stationary point. By finding $R'(\theta)$ and
setting it equal to zero, we find the stationary point[*4]

$$\cos\theta = \left(\frac{r_2}{r_1}\right)^4, \qquad \text{or} \qquad \theta = \arccos\left(\frac{r_2}{r_1}\right)^4, \qquad (45)$$

where the function $\arccos(y)$ returns the angle whose cosine is the
number y. As an example of this $\cos\theta$ equation, if r_2 is 75% the size
of r_1, then $\theta \approx 71.5°$.

We've just extracted an important insight from Poiseuille's law: the
branching angle that minimizes the resistance depends on the ratio of
the radii of the vessels at the branching point. Imagine yourself now as
a developing baby in your mother's womb. As your tiny body begins to

grow the vast number of blood vessels in your body begin to branch. As your thicker vessels—such as your arteries—branch off into smaller vessels, the optimal branching angle changes based on the ratio r_2/r_1. And throughout the millions of years that we've evolved, our bodies have been constantly adjusting these branching angles in an attempt to minimize the energy expended by the heart in circulating blood.

I find it truly amazing that we've been able to understand how biology uses optimization to make our bodies more efficient. But minimizing the energy needed to accomplish something isn't only a feature of biological systems. For example, take the power lines I see outside my window. They hang in a particular shape that one might think has nothing to do with optimization. But in fact, as Newton helped us understand, all objects on Earth are pulled downward by Earth's gravity; this includes the power line too. If we think of the power line as a collection of tiny pieces glued together, every piece of the cable wants to be on the ground; but since all the pieces are connected, the best they can do is to *minimize* their distance to the ground. The shape that emerges from this tug-of-war is called a *catenary* (Figure 5.4).

Although this shape looks like a parabola, it's actually not. This is evident from the catenary's equation,

$$y = a \cosh\left(\frac{x}{a}\right), \tag{46}$$

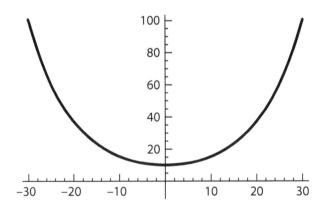

Figure 5.4. The graph of a catenary.

where $a \neq 0$ is a number and $\cosh(x)$ is the hyperbolic cosine function. In this shape, the power line minimizes its gravitational "stored energy," much like a ball close to the ground has less gravitational stored energy than one farther off the ground.

The tendency of Nature to prefer minimum energy configurations was made precise by the Irish physicist and mathematician William Hamilton. In 1827 he presented what we today call *Hamilton's Principle of Stationary Action* to the prestigious Royal Society of London. The fact that it has the word "stationary" should be a clue as to what this principle says: among all possible trajectories that a physical system could take between two states A and B, the one it actually takes renders the action S stationary. And for a wide variety of physical systems— the hanging power line included—a stationary point for S corresponds to a minimum energy configuration.[xxi] So, as you look around Nature, whether it's hanging power lines, water flowing down a stream, or planets orbiting a sun, Hamilton's principle tells us that Nature is, behind the scenes, optimizing it all.

The Costly Downside of Calculus

My admiration of the optimization going on all around us is interrupted by my phone. It's Zoraida; she's calling to see what we should do after work. Since it's a Friday, both of us mention going to downtown Boston. We settle on dinner and a movie, and the plan is for me to meet her downtown in about an hour and a half. I'll need to get home first and then take the T (our light rail system) downtown. To save us some time, I decide I'll buy the movie tickets online now. The theater's website has the tickets going for $12. How come they aren't $15, or $20? Seems like a lot, which gets me thinking: would the theater make more money at those higher prices? This leads to the more general question: how should a theater—or any business for that matter—set its prices?

[xxi] Although a more precise mathematical statement of Hamilton's principle involves a field of mathematics known as the calculus of variations, the condition that S be stationary reduces to the condition that the appropriately defined derivative of S be zero.

Although this is a complicated question, let's focus on the theater's revenue and ask: what should the ticket price be to maximize revenue?

We'll need some initial assumptions. Let's denote by p the price of the theater's movie ticket and assume the theater seats 2,000 people in total throughout all of its screen rooms. Suppose that with ticket prices at \$12 the average attendance last month was 1,000. Businesses often conduct surveys on hypothetical price changes to try to estimate changes in demand for their products. Suppose that the theater conducted such a survey and discovered that for every 10 cents the ticket price dropped the theater would attract another 20 moviegoers. Using all this information, we can now determine the ticket price that maximizes the revenue from ticket sales. Here's how.

From the survey we know that after one 10-cent decrease, the price is $p = 12 - (1/10)$, and after two 10-cent decreases the price is $p = 12 - (2/10)$. Therefore, if x represents the number of 10-cent decreases in the price of the original \$12 ticket, after x 10-cent decreases the price would be

$$p = 12 - \frac{x}{10}. \qquad (47)$$

The survey also implies that after one 10-cent decrease the average attendance would rise to $1,000 + 20$, and after two 10-cent decreases it would rise to $1,000+20(2)$. So after x 10-cent decreases the average attendance would rise to

$$1,000 + 20x. \qquad (48)$$

Since the ticket revenue R is the total attendance multiplied by the ticket price, we now know that

$$R(x) = (1,000 + 20x)\left(12 - \frac{x}{10}\right) = 12,000 + 140x - 2x^2. \quad (49)$$

What about the interval? Well, the theater could decrease the price zero times, or it could decrease the price 120 times (at which point the ticket would cost \$0, leading to zero revenue). Thus our interval is $0 \leq x \leq 120$. But since $R(120) = 0$ we know this isn't the maximum revenue. On the other hand, $R(0) = \$12,000$, which *could* be the maximum revenue; in this scenario the theater should keep prices at $12. But we've yet to check the stationary points. Calculating $R'(x)$ yields[*5]

$$R'(x) = 140 - 4x. \tag{50}$$

From this we see that the only stationary point is $x = 35$ (since $R'(35) = 0$). But remember, the theater was already generating $12,000 by selling $12 tickets, so we need to see if reducing the price by 35 10-cent chunks will result in higher revenue. Well, from our revenue function we find $R(35) = \$14,450$,[*6] the largest revenue of the stationary points and the endpoints of the interval; hence this is our maximum revenue.

These 35 10-cent reductions amount to a $3.50 discount to the current price, bringing each ticket's price down to $8.50. And, since the theater determined it would bring in an additional 20 moviegoers per reduction, at 35 reductions this would represent an increase in average attendance of 700 people. If only I could call up the theater and get a discount by offering to use calculus to maximize its revenue from ticket sales. Unfortunately for me, my computer screen still shows $12. I guess that's the price of seeing a movie these days.

The Optimal Drive Back Home

After buying the tickets and making dinner reservations at an Indian restaurant (Zoraida's favorite), I pack up and head down the three flights of stairs for the last time today. Walking to my car, I can't shake the feeling of having paid too much for those movie tickets. I guess that's a downside of optimization: businesses use it to maximize their profits, and hence our costs. But then I realize that *we* can also think like a business and use calculus to lower our own costs.

As I start my car and see the gas gauge rise I spot my first target for cost reduction: fuel costs. I usually follow the same route back home

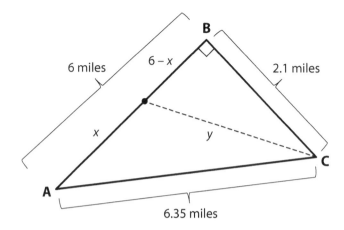

Figure 5.5. An illustration of the routes I can take to get home from work.

from work, but today I'll do something different. My goal for this trip is to follow the route that minimizes the number of gallons of fuel used. With my seat belt on and 20 minutes left before I get home I start my mental gears turning, determined to compensate for my feeling of having overpaid for the tickets. This time *I* will be the one using calculus to my advantage.

When I leave work I have a few routes that I can take (Figure 5.5). Between my work (labeled *A*) and my house (labeled *C*) there are several connecting roads. The road connecting points *A* and *B* has a maximum speed limit of 50 mph, and all other roads connecting this high-speed road to my house are city roads with a maximum speed limit of 30 mph. The question is: how far should I drive down the high–speed road before taking a side road in order to minimize the amount of fuel used for the trip?

The first ingredient is my car's fuel economy. My car gets 36 mi/gal on highways and 29 mi/gal on city roads. So if I drove down the high-speed road for *x* miles and then took a side road the remaining distance *y*, the total gallons *g* of fuel used would be

$$g = \frac{x}{36} + \frac{y}{29}. \tag{51}$$

By using the distances between the points A, B, C (Figure 5.5), we can express g as a function of x:[*7]

$$g(x) = \frac{x}{36} + \frac{\sqrt{(6-x)^2 + 4.41}}{29}. \tag{52}$$

In this case finding the endpoints is easy: I could choose to drive on the high-speed road all the way to the point B, or I could choose to never take the high-speed road and instead drive from A to C directly. When expressed in terms of x, this gives the interval $0 \leq x \leq 6$. And since $g(0) \approx 0.22$ gallon and $g(6) \approx 0.24$ gallon, unless the stationary points give a $g(x)$–value less than 0.17 I'll be staying on the high-speed road all of the way to the point B.

Finding $g'(x)$ and setting it equal to zero we find[*8] that the only stationary point inside the interval $0 \leq x \leq 6$ is $x \approx 3.14$ miles. And since $g(3.14) \approx 0.21$, the optimal route is to take the high-speed road for just over 3 miles and then take a side road to the point B.

Now that I'm taking the optimal route back home I can enjoy the rest of the trip, knowing I'm using the minimum amount of fuel to get there.[xxii] It might seem silly to have done all this work since the difference between the minimum fuel and maximum fuel routes is just $0.24 - 0.21 = 0.03$ gallon. At \$4 per gallon, this saves me a *measly* 12 cents. But imagine now that you use these same optimization techniques to save UPS or FedEx 20 *measly* cents in fuel costs for every 8.1 miles (the total distance I'll travel home) that *each* of their trucks travel. With the large bonus you'd be paid you could probably retire immediately!

Catching Speeders Efficiently with Calculus

About halfway home, while still on the high-speed road, I spot a police car in the distance. I'm traveling just under the speed limit, which

---

[xxii] Technically, even this slightly complicated analysis is a drastically simplified treatment of this problem, since a lot of factors have been neglected. For example, we know that fuel economy depends on velocity. Moreover, the roads connecting the points A, B, C are not as straight as I've made them seem. But hey, you've got to start somewhere, right?

means that the cars passing me are probably speeding. However, if
the cop's radar gun doesn't clock the driver's speed at exactly the
right time, he or she might dodge the speeding ticket. And since
most of us slow down when we see a police car anyway, this driver
will likely notice this and avoid the speeding ticket. These facts show
that this method of catching speeders is very inefficient; if the police
department is trying to maximize its revenue from speeding tickets, or
minimize the accidents caused by speeding, there has to be a better
way to catch speeders. Lucky for the cops, there is. And although
it only indirectly involves optimization, it has everything to do with
one of the most important theorems in calculus: the Mean Value
Theorem.

Put simply, the Mean Value Theorem (MVT) says that if you draw a
differentiable function $f(x)$—a continuous function that is "smooth"—
and draw a line between any two points $(a, f(a))$ and $(b, f(b))$ on the
graph, then there is some point c between a and b for which the slope of
the tangent line at c, $f'(c)$, is the same as the slope of the line through
the two points you drew. "What?!" This is one of those times when a
picture really is worth a thousand words, so have a look at Figure 5.6. By
comparing Figure 5.6 with the statement of the Mean Value Theorem

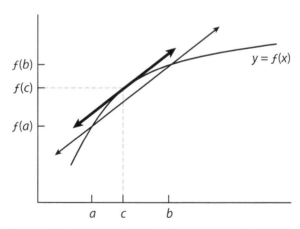

Figure 5.6. An example of the Mean Value Theorem at work. The heavier line has
slope $f'(c)$.

we see that the main nugget of the MVT is the existence of an x-value
c, where $a < c < b$, for which

$$f'(c) = \frac{f(b) - f(a)}{b - a}. \tag{53}$$

At this point you may be asking yourself how in the world this math
theorem can help catch speeders. The answer is another excellent
example of how calculus connects seemingly unrelated things.

Imagine that the cop I see off in the distance clocks the driver
that just passed me. But because the driver saw the police car and
slowed down in time, the cop's radar gun shows the driver's speed
to be 50 mph, exactly the speed limit. But pretend now that one mile
down the road and 50 seconds later, another police car clocks the same
driver going 45 mph. Although the driver wasn't speeding at both of the
instants the two cops measured his speed, if these two cops compared
their notes—and used the Mean Value Theorem—they'd conclude that
at some time c inside the interval $0 < t < 50$ the driver's velocity $s'(c)$
must have been equal to

$$s'(c) = \frac{s(50) - s(0)}{50 - 0} = \frac{1 \text{ mile}}{50 \text{ seconds}} = 72 \text{ mph}.$$

On this 50 mph road, that would definitely qualify as speeding! And
believe it or not, some highways actually use this basic idea—that of
comparing the speeds of cars at the start and end of a fixed distance
along a highway—in an automated way using cameras; so the next time
you see two tall posts with cameras separated by a hundred feet or so
along a highway, slow down!

Aside from thinking about ways to help cops catch speeders by
using the MVT, the rest of my drive home is uneventful. Though
I saved only 20 cents by taking the optimal route, I arrive at my
house happy that I've made several connections between the seemingly unrelated fields of biology, business, physics, and "saving money
when driving home" through the mathematics of optimization. What
I find particularly satisfying is that the notion of a stationary point,

and its central importance in optimization, emerged out of the simple analysis of the trajectory of my hot chocolate cup. Sometimes the most profound insights come from thinking very carefully about the implications hidden behind the simplest of phenomena. And as we'll see in the next chapter, this seemingly silly MVT example forms the foundation of calculus's other half: the theory of integration.

CHAPTER 6

ADDING THINGS UP, THE CALCULUS WAY

MY DRIVE HOME TOOK ME about 20 minutes. I'm making a quick pit stop inside my house to change out of my work clothes. For me, changing into "non–work clothes" basically means just changing into jeans. And since I own only four pairs of jeans, five quick minutes later I'm changed and ready to catch the T.

I never used public transportation until I moved up north, having grown up in Miami, Florida. But since our home is a one-minute walk from the nearest T station, it's a no-brainer. The Massachusetts Bay Transportation Authority (MBTA) operates the T; in fact, Boston is home to the oldest subway tunnel in the United States, dating back to 1897.[23] With such a long history of ridership it's no surprise that the MBTA's rail system is widely used today. In 2009 the MBTA system ranked fifth in the nation in overall ridership, completing about 370,000 passenger trips for a total of 1.8 *million* miles *in that year alone.*[24]

With so many trains to keep track of and such a high demand for its services, the MBTA needs to continuously determine the best time to take trains out of service for maintenance. This sounds like an optimization problem, but the MBTA uses the simpler approach of taking trains out of service after they've traveled a certain number of miles, much like we change the oil in our cars every 3,000 to 5,000 miles. There's one problem with this: how do we calculate the distance a particular rail car has traveled? If it traveled in a straight line this would be easy. But since the tracks twist and curve, we need a way of adding up all of those distances. Notice that this problem is fundamentally different from those we've been studying. This isn't a question about *change*, so we don't expect to find derivatives lying around. It seems that we don't have anything to start from. But don't worry, since in the 40 minutes I'll need to get downtown I'll have plenty of time to tell you about differentiation's twin brother: integration.

The Little Engine That Could . . . Integrate

The train stop I'm at looks like any other train stop you could imagine. There are a couple of enclosed seating areas to keep us all warm in winter and machines that you can use to buy tickets. Then there are the rails, of course. They continue as far as you can see in both directions. And somewhere off in the distance I spot a faint outline of what looks to be the rail car. The line I'll be hopping on is the green D line,[xxiii] and it's the fastest of the green-line trains. Since the train is still far away from the platform, the conductor can operate the train at a relatively high velocity. From my experience riding the train, it's likely she's running the train at a constant velocity, I'd say about 35 mph. I know the train will reach me in the next few minutes, which gives me the opportunity to discuss the distance problem we started the chapter with. Let me ask an easier question: if the train is traveling at 35 mph, can I determine how far away it is?

The short answer is yes. I could simply use the formula "distance equals rate times time." My best estimate is that the train will reach me in about 30 seconds—or about 0.0083 hour—so that my estimate for the distance d (in miles) the rail car is away from me is

$$d = 35(0.0083) \approx 0.3 \text{ mile.} \tag{54}$$

Let's try to visualize this answer with a graph. If we denote by $v(t)$ the velocity of the train, then $v(t) = 35$ (since I've assumed it to be moving at the constant velocity of 35 mph). Figure 6.1(a) shows the graph of this function, but how does it show that the distance traveled is 0.3 mile?

[xxiii] The MBTA rail system is divided into different lines grouped into colors based on the regions they serve. The green line runs east to west and back. Some of these lines, like the green line, split into other lines (denoted by letters) that serve specific destinations.

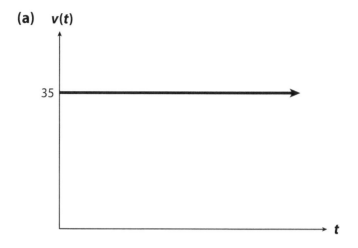

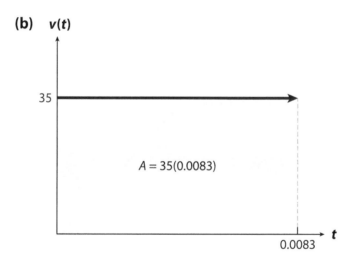

Figure 6.1. (a) The graph of the function $v(t) = 35$. (b) The area of the shaded region is the distance traveled in 0.0083 hour.

Well, the geometric equivalent of calculating distance as rate times time ($d = rt$ for short) is finding the *area* of the rectangle in Figure 6.1(b). This is another one of these simple insights that will turn out to have *profound* consequences.

But the answer for d in (54) is not quite correct. Remember, we assumed the train was traveling at the constant velocity of 35 mph;

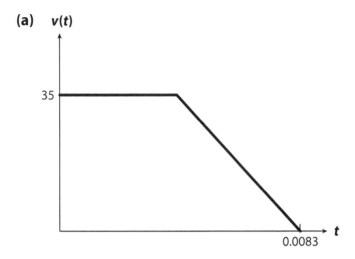

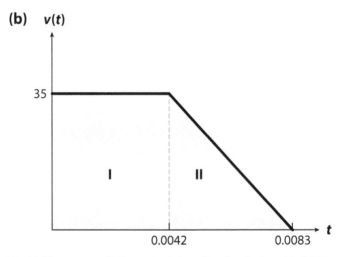

Figure 6.2. (a) The more realistic graph of the rail car's velocity $v(t)$. (b) The sum of the areas of the shaded regions is the distance traveled.

but our calculation also tacitly assumes this to be true *right up to the instant the rail car reaches me*, at which point it stops *instantaneously*. In real life this would be dangerous for everyone on the train, so we know that sometime before the conductor approached the platform the train began to slow down. Figure 6.2(a) shows a more realistic velocity function. There I've assumed the conductor started slowing down (with

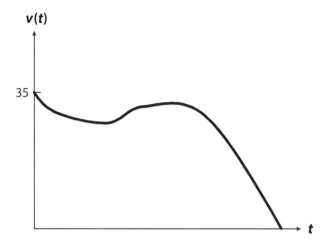

Figure 6.3. An even more realistic graph of the rail car's velocity $v(t)$.

a constant deceleration) when the train was 15 seconds, or 0.0042 hour, away from the platform.

To find the new distance traveled we can still use $d = rt$, except now we need to find two different areas: that of a rectangle (region I in Figure 6.2(b)), and a triangle (region II in Figure 6.2(b)). The two areas correspond to the two distances d_1 and d_2, and the sum $d_1 + d_2$ is the total distance the train is away from me. By calculating the areas, this new estimate gives about 0.22 mile.[*1]

This is certainly a better approach, but yet again we've made the restrictive assumption that the conductor is decelerating at a constant rate. What if this rate isn't constant? And what if the velocity of the train before decelerating isn't *exactly* constant at 35 mph? Figure 6.3 shows an even more realistic velocity function (last one, I promise) that would account for these factors.

We can try to follow the same prescription to find the distance traveled (find the area under the graph of $v(t)$), but we don't yet know how to find areas of curved regions. We've run into one of the classic problems that stumped mathematicians for *two millennia*. But if there's one thing that calculus has taught us, it's that we can work miracles by taking limits. After all, we defined the derivative $f'(x)$ in Chapter 2 in terms of the limit as h approached zero of a slope, and in Chapter 5

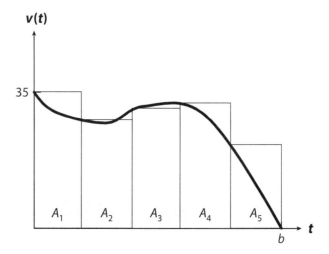

Figure 6.4. Approximating the area under the curve by using five rectangles.

we introduced differentials as limits of the changes Δf and Δx. This approach suggests we take the limit of the quantity we've been working with thus far: *area*.

Let's first use areas of rectangles to *approximate* the area A under the graph of $v(t)$ in Figure 6.3. Let's denote the t-value where $v(t) = 0$ by b, and use five rectangles for our approximation (see Figure 6.4). We don't know the answer yet, but let's denote it by

$$\int_0^b v(t)\, dt. \tag{55}$$

The elongated S in this expression is called an *integral* sign, and the dt keeps track of the independent variable we're using (t in our case). Notice also that the 0 and b keep track of the fact that the area we're finding extends from $t = 0$ to $t = b$. For these reasons, this funny notation would be read out loud as "the definite integral of the function $v(t)$ with respect to t from $t = 0$ to $t = b$," and this is how we mathematicians denote the area under the curve; the adjective "definite" is used whenever the "limits of integration," in this case 0 and b, appear along with the integral sign.[xxiv] Since we're estimating the true

---

[xxiv] If the limits of integration were absent, we'd call this an *indefinite* integral.

area by using the area of the five rectangles in Figure 6.4, mathematically we're saying that

$$\int_0^b v(t)\,dt \approx A_1 + A_2 + A_3 + A_4 + A_5. \tag{56}$$

To keep this simple, we've assumed that the rectangles all have the same width, in this case $b/5$, and that the upper-left vertex of each one touches the graph of $v(t)$ and is the rectangle's height. For example, the first rectangle has height $v(0)$, the second $v(b/5)$, the third $v(2b/5)$, and so on, with the last rectangle's height equal to $v(4b/5)$. Since the area of a rectangle is its width times its height, our estimate becomes

$$\int_0^b v(t)\,dt \approx \frac{b}{5}v(0) + \frac{b}{5}v\left(\frac{b}{5}\right) + \frac{b}{5}v\left(\frac{2b}{5}\right) + \frac{b}{5}v\left(\frac{3b}{5}\right) + \frac{b}{5}v\left(\frac{4b}{5}\right)$$

$$= \frac{b}{5}\left[v(0) + v\left(\frac{b}{5}\right) + v\left(\frac{2b}{5}\right) + v\left(\frac{3b}{5}\right) + v\left(\frac{4b}{5}\right)\right]. \tag{57}$$

Now, even though we used only 5 rectangles, we could just as easily have used 10, 100, or any other number n. If we insist that they all have the same width, then each rectangle would now have width b/n. Extrapolating the pattern above, the first rectangle would still have height $v(0)$, while the second would have height $v(b/n)$, the third $v(2b/n)$, and so on, until we reach the last rectangle, whose height would be $v((n-1)b/n)$. Our new estimate would be

$$\int_0^b v(t)\,dt \approx \frac{b}{n}\left[v(0) + v\left(\frac{b}{n}\right) + \cdots + v\left(\frac{(n-1)b}{n}\right)\right]. \tag{58}$$

This sum of areas is called a *Riemann sum* after the German mathematician Bernhard Riemann, who rigorously solved the "area under the curve problem" in 1853. Mathematicians use a shorthand notation for the sum in the brackets:

$$\sum_{i=0}^{n-1} v\left(\frac{ib}{n}\right) = v(0) + v\left(\frac{b}{n}\right) + \cdots + v\left(\frac{(n-1)b}{n}\right). \qquad (59)$$

The E-looking symbol on the left is the Greek letter sigma, and the expression on the left-hand side of the equation is an instruction: add together the quantities $v(ib/n)$ starting with $i = 0$ and ending with $i = n - 1$. In this notation our estimate becomes

$$\int_0^b v(t)\, dt \approx \frac{b}{n}\sum_{i=0}^{n-1} v\left(\frac{ib}{n}\right) = \sum_{i=0}^{n-1} v\left(\frac{ib}{n}\right)\frac{b}{n}. \qquad (60)$$

We have now gone as far as we can go without calculus, but luckily only one more intuitive step is required. When looking back at Figure 6.4, it's apparent that using five rectangles in our approximation is better than using one. This suggests that we get better approximations when we increase the number of rectangles. The logical conclusion: if we could use an *infinite* number of rectangles, we'd get the *exact* area instead of an approximation. In calculus-speak, this suggests we take the limit of the Riemann sum as the number of rectangles n approaches infinity. What we get is

$$\int_0^b v(t)\, dt = \lim_{n\to\infty}\sum_{i=0}^{n-1} v\left(\frac{ib}{n}\right)\frac{b}{n} = \lim_{n\to\infty}\sum_{i=0}^{n-1} v(t_i)\frac{b}{n}, \qquad (61)$$

where in the last equation I've written t_i for ib/n to highlight the following important fact: although we found the height of each rectangle from the condition that the upper-left vertex of each rectangle touch the graph, we could have used the upper-*right* vertex, or even the midpoint of the rectangle. In fact, if we denote by t_i the t-value at which a rectangle touches the graph, then $v(t_i)$ would be the height of that rectangle (see Figure 6.5).

Whew! That took a lot of work! But let's step back and see what we've accomplished. Given a positive velocity function $v(t)$ (like the one in Figure 6.3), formula (61) allows us to find the total distance

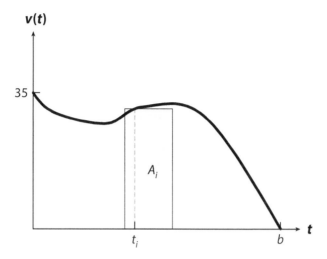

Figure 6.5. A close-up of a rectangle with height $v(t_i)$.

traveled by calculating areas, adding them up, and taking a limit. This is pretty incredible, since not only did we solve in a few pages a problem that stumped mathematicians for thousands of years, but what we've done can be applied to *many* other functions. For example, for any continuous function $f(x)$, the area under its graph between $x = a$ and $x = b$,

$$\int_a^b f(x)\, dx, \tag{62}$$

can be found using exactly the same limiting argument we just developed. This other half of calculus dedicated to studying the integrals of functions is called *integration*, and along with the study of differentiation comprises the two pillars of calculus. And like our adage "wherever there is change, derivatives can be found," the Riemann sum foundation of the definite integral allows us to state a parallel mantra: "whenever quantities need to be added together, integrals are not far behind."

But there's one drawback to our formula (61): in practice it's really hard to compute an integral by calculating limits. We'd like to find a shortcut, much like we did when we moved away from using limit tables in Chapter 2 to calculate the derivative. And in a nice twist of fate, although derivatives deal with slopes of tangent lines and integrals have to do with areas under the curve, the two subjects are related in the most *beautiful* way that will also turn out to resolve our computational difficulties.

The Fundamental Theorem of Calculus

As I knew it would, the rail car arrived safely at the platform, and I board the train. Luckily there are a few open seats (not always the case), and I plop down and pull out my phone. I text Zoraida that I'm on my way, and that she should meet me at the Indian restaurant in about 30 minutes, about the time I expect the train to take to get me downtown. Now that that's done I can get back to the business of calculating the integral.

The way we approached it involves a lot of steps. First find the quantity $v(ib/n)$; then calculate the Riemann sum; finally, take the limit and arrive at the answer. There *has to* be an easier way.

By now you've probably anticipated that there is. What you may not have anticipated is where this easier method will come from.

The fundamental link between the subjects of integration

and differentiation boils down to our ability to catch speeders.

Allow me to explain this surprising turn of events.

Remember that in Chapter 5 we imagined two cops communicating with each other, using the Mean Value Theorem (MVT) to catch speeders. Using this theorem they could conclude that, between any two points in time $t = a$ and $t = b$, there is a time c for which the driver's velocity $v(c)$ satisfies the equation

$$v(c) = \frac{s(b) - s(a)}{b - a}, \tag{63}$$

where $s(t)$ is the driver's position function. Now replace the driver and the car with the conductor and the rail car, and you realize that we could just as easily apply the MVT to our moving train.

To find out what happens, let's concentrate on the first interval, $0 \leq t \leq b/n$. The MVT says that there is some t-value, let's call it t_0, in the interval $0 \leq t \leq b/n$ at which the rail car's velocity $v(t_0)$ satisfies

$$v(t_0) = \frac{s\left(\frac{b}{n}\right) - s(0)}{\frac{b}{n} - 0}, \quad \text{or} \quad \frac{b}{n}v(t_0) = s\left(\frac{b}{n}\right) - s(0). \tag{64}$$

But we can also apply the MVT on the interval $b/n \leq t \leq 2b/n$, and also on $2b/n \leq t \leq 3b/n$, and so on. The results would be the additional intermediate t-values $t_1, t_2, \ldots, t_{n-1}$, along with the relationships

$$\frac{b}{n}v(t_1) = s\left(\frac{2b}{n}\right) - s\left(\frac{b}{n}\right), \quad \cdots, \quad \frac{b}{n}v(t_{n-1}) = s(b) - s\left(\frac{(n-1)b}{n}\right). \tag{65}$$

If we now add up all of these equations, we obtain[*2]

$$\sum_{i=0}^{n-1} v(t_i)\frac{b}{n} = s(b) - s(0). \tag{66}$$

Using this result in our formula for the distance traveled by the rail car gives

$$\int_0^b v(t)\,dt = \lim_{n\to\infty}\left[\sum_{i=0}^{n-1} v(t_i)\frac{b}{n}\right] = \lim_{n\to\infty}[s(b) - s(0)] = s(b) - s(0), \tag{67}$$

since the quantity $s(b) - s(0)$ is a number that doesn't change as n approaches infinity.

What we've just discovered is a *much* more useful way to calculate the integral of our $v(t)$ function over the interval $0 \leq t \leq b$. Formula (67) says that the answer is just $s(b)$ (the rail car's position at time $t = b$) minus $s(0)$ (the rail car's position at time $t = 0$). In other words, the distance the rail car travels in getting to the platform, $s(b) - s(0)$, is found by integrating the rail car's velocity function $v(t)$ from $t = 0$ to $t = b$.

Put this way our result doesn't seem all that "fundamental"; it seems to tell us something we already knew: the distance traveled is the difference between where the rail car ended ($s(b)$) and where it started ($s(0)$). But let's rephrase what we found for a general function $f(x)$:

$$\int_a^b f(x)\,dx = F(b) - F(a). \tag{68}$$

Question: What is F? Well, in our velocity example F was the distance function $s(t)$, whose *derivative* $s'(t)$ is the velocity function $v(t)$. This turns out to be the relationship in the general case as well, so that F is related to f by

$$F'(x) = f(x), \qquad \text{or} \qquad F(x) = \int f(x)\,dx. \tag{69}$$

Mathematicians call the function $F(x)$ the *antiderivative* of $f(x)$; it's the function that, when differentiated, gives $f(x)$.[xxv] Therefore, here is the easier way to calculate the definite integral I had promised earlier. To tie it all together, formula (68) says that to find the definite integral of $f(x)$, we should first find an antiderivative $F(x)$. Then, simply calculate $F(b) - F(a)$.

To illustrate how powerful this new approach is, let's use it to confirm one of the facts we talked about in Chapter 1: that anything you throw up in the air follows a parabolic trajectory.

Let's start with Galileo's realization that all objects fall at the same constant acceleration $a(t) = -g$. Since an object's velocity $v(t)$ is related to its acceleration $a(t)$ by $v'(t) = a(t)$, in our new terminology $v(t)$ is the antiderivative of $a(t)$, and[*3]

$$v(t) = \int a(t)\,dt = \int -g\,dt = v_0 - gt, \tag{70}$$

where v_0 is the object's initial velocity. Furthermore, since we know that the object's position—let's focus on its vertical position $y(t)$ for now—is

---

[xxv]Notice that the integral sign is also used, and in this case it's referred to as the *indefinite* integral.

related to its velocity $v(t)$ by $y'(t) = v(t)$, then[*4]

$$y(t) = \int v(t)\,dt = y_0 + v_0 t - \frac{1}{2}gt^2, \tag{71}$$

where y_0 is the object's initial vertical position. This rederives the formula we used in Chapter 1 when analyzing the vertical position of a water droplet coming out of my showerhead, but this time we did it without the factoid we used about objects moving with speeds that vary linearly with time.

The fact that formula (68) relates the *integral* of the function $f(x)$ to its *antiderivative* $F(x)$ means that it connects the two pillars of calculus—integration and differentiation. In fact, if we substitute $f(x) = F'(x)$ into formula (68), we get

$$\int_a^b F'(x) = F(b) - F(a), \tag{72}$$

telling us that integration and differentiation *undo each other*. For these reasons, formula (68) is referred to as the Fundamental Theorem of Calculus. But despite its important-sounding name, remember that at the very heart of this powerful conclusion stood our old speeder-catching friend, the Mean Value Theorem.

Using Integrals to Estimate Wait Times

I've spent the train ride thus far marveling at the deep connections between the MVT, integration, and differentiation. When I look outside the window I notice that the train has stopped. It does so periodically (the rail car obeys certain signals to prevent it from crashing into the one in front of it), but it's now been stopped for longer than usual. The conductor starts informing us through overhead speakers that the rail car in front of us has broken down. Great; I'm only five minutes away from my stop and now I'm stuck here for who knows how long. I guess this is one of the downsides of having the oldest subway system in the country.

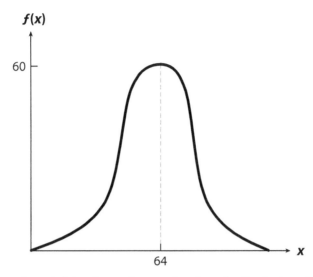

Figure 6.6. The probability density function describing the distribution of heights for adult women in the United States.

My first thought is to call Zoraida to let her know I'll be late, but my phone doesn't get reception here. I'm just going to have to wait, but for how long? If I'm stuck here for more than five minutes I'll be late to the rendezvous with Zoraida. So the more relevant question is: how likely is it that I'll wait more than five minutes? This is a question about *probability.*

We're all somewhat familiar with probability. Imagine a bag containing three red marbles and seven blue ones. If you reach in and pick out a marble, what's the probability it'll be a red one? The answer is 3/10, or 30%. And since all probabilities must add up to one, the probability of picking a blue marble is $1 - 0.3 = 0.7$, or 70%.

But if the number of outcomes is a continuous variable x, then things get a bit more complicated. In these cases, there is a function $f(x)$ called the *probability density function* (PDF). For example, the PDF describing the distribution of heights for adult women in the United States is called a *Gaussian* distribution, and its graph is the familiar "bell-shaped" curve (Figure 6.6). This curve is obtained by sampling a large number of adult women in the United States and recording how common a certain height is. For example, the graph shows that 60% of the sample has a height of 64 inches. The fact that this is the most frequent height among adult women in the United States makes this the average height for that population.[25]

Now, the interpretation for the PDF is simple: it describes the probability that x takes on a certain value. Therefore, if we'd like to find the probability that x lies in some range $a \leq x \leq b$, denoted by $P(a \leq x \leq b)$, we need to *add up* the probabilities that x takes on *each* value in the interval from a to b. According to our new mantra, there is an integral lurking around somewhere. Indeed, the same argument used to arrive at formula (67) leads us to

$$P(a \leq x \leq b) = \int_a^b f(x)\,dx. \tag{73}$$

But I'm interested in the probability of waiting *more than* five minutes. Since all probabilities add to one, this is equivalent to one minus the probability of waiting *less than* five minutes, or

$$P(x > 5) = 1 - \int_0^5 f(x)\,dx. \tag{74}$$

When it comes to waiting times, a common PDF that's used is the exponential distribution

$$f(x) = \frac{1}{m}e^{-x/m}, \quad x \geq 0. \tag{75}$$

Here m is the average wait time; if I knew m in my case, I could use this PDF to calculate the probability that I'll be waiting for more than five minutes. Based on my past experience, I've never had to wait more than ten minutes for a train. And sometimes I've arrived at the stop just as the train is pulling up; so my best estimate for m is five minutes. Assuming that this is also a reasonable average wait time for fixing disabled trains, let's set $m = 5$. Using this we can now use the Fundamental Theorem of Calculus to get[*5]

$$1 - \int_0^5 \frac{1}{5}e^{-t/5}\,dt \approx 0.368, \tag{76}$$

or about 36.8%. That's great, since it means I'm unlikely to be stuck here for more than five minutes. Sure enough, a short minute later the conductor informs us that the rail car in front of us is now moving. It looks like I won't be late after all.

In the last few minutes I have left on the train, let me point out that this seemingly simple application of integration has real uses in the business world. For example, it's important for companies to have information on wait times. We've all experienced what happens when this *isn't* something a company cares about. The ensuing long wait times are often frustrating for the customer, in effect acting as a disincentive to buy the company's products. Companies can now combat this by determining their own PDFs and using integration to calculate the probabilities of different wait times. The results could then help the management team determine what aspects of the company—such as customer call centers and shipping warehouses—need improved wait times.

In the last chapter we talked about how a company might use derivatives to optimize its revenue or profit streams. The mathematics of this chapter has taught us that *integrals* are just as important a tool for this same purpose. And now that we know that calculating probabilities boils down to integrating PDFs, this opens up a whole new world of possible applications of integration. Essentially *anything* involving a heavy dose of probability (for example, sports) now benefits from the integration mathematics we've developed. But there's *much* more we can do with integration. Although it's harder to think about, many instances call for adding up an infinite number of quantities. And according to our new mantra, in each of these instances integrals can be found. In the next chapter we'll combine what we know about derivatives and integrals to tackle some of the biggest questions in all of human existence.

DERIVATIVES *AND* INTEGRALS: THE DREAM TEAM

AFTER FIRST DISCUSSING derivatives in Chapter 2 we not only found them everywhere, but as our discussion of time dilation showed, they were powerful enough to literally change our view of reality. Similarly, after introducing integration in the last chapter we already have seen how it naturally appears all around us via probability. Imagine, then, the possibilities of using both derivatives and integrals in mathematizing a situation. As we'll soon find out, this dream team can answer some of the most fundamental questions posed in the history of civilization. But I'll have to build up to that crescendo first. So let's get back to the rendezvous with Zoraida.

When you walk out of the Boylston station in downtown Boston you see a sign that says "Boston Common, Founded 1634." As the oldest public city park in the United States, Boston Common's 50 acres of land was at one time used as a camp by the British before the American Revolutionary War.[26] Today it's one of the main congregation points for residents of the greater Boston area. Next to the Common is the Public Garden. In the summer months you'll find a plethora of colors here, coming from the variety of plants and flowers scattered throughout the grounds. The whole scene makes for a nice welcome to downtown Boston, and invites us to reflect on all of the history the city has to offer. I soak up the scenery on my way to the Indian restaurant, which is only a few blocks from the Boylston stop.

Integration at Work—Tandoori Chicken

I'm right on time, and Zoraida and I are seated by the window. Looking through the menu, I spot my favorite tandoori chicken. This roasted chicken dish is marinated in yogurt and spices, and seasoned with tandoori masala powder. It's traditionally cooked in a tandoor oven—a bell-shaped clay oven that can reach temperatures of up to 900° F. Traditionally this high heat was generated by using charcoal or wood, but these days it's probably generated by electricity or natural gas.

From what other Indian cuisine lovers have told me, the tandoori chicken is cooked at about 500°F. And however the heat is generated, the goal is to keep the oven at 500° F. Nowadays ovens do this with built-in *thermostats*. These nifty little devices maintain a prescribed average temperature by cycling on and off according to the temperature its built-in thermometer measures. But since the temperature inside the oven might be different from nanosecond to nanosecond, what exactly do we mean by the "average" temperature? And how does the oven actually maintain this temperature? (Sorry, I can't help it, I see math *everywhere*.)

The first thing to realize is that the restaurant's tandoor oven is likely already preheated, since other diners have probably been ordering dishes that require its use. For simplicity, let's assume that when Zoraida and I walked into the restaurant its temperature was 525° F, and that the oven thermostat is set to 500° F. Having reached this level, the thermostat would've turned off the heat,[xxvi] and the oven would have started to cool down. Once the oven reached, say, 475° F, the thermostat would kick in and turn on the heat. The ensuing temperature $T(t)$ in the oven therefore looks something like the graph in Figure 7.1. The question is: how do we express mathematically the requirement that the average temperature be 500° F? One clue comes from the word "average."

We all know a little something about averages. If you imagine three people who are 60, 65, and 70 inches tall, then their average height

---

[xxvi]Oven manufacturers program ovens to cycle on and off once the temperature has increased or dropped by a certain amount relative to the temperature the user has requested.

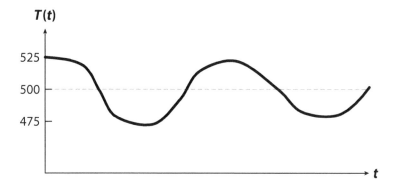

Figure 7.1. A reasonable graph of the temperature $T(t)$ inside the tandoor oven.

would be

$$\frac{60 + 65 + 70}{3} = 65 \text{ inches.} \tag{77}$$

If we created the function $f(x)$ to record the height of person x and denoted the height of person 1 by x_1, person 2 by x_2, and so on up to x_n, then the average height would be

$$f_{\text{avg}} = \frac{f(x_1) + f(x_2) + \cdots + f(x_n)}{n}, \quad \text{or} \quad f_{\text{avg}} = \frac{1}{n}\left[\sum_{i=1}^{n} f(x_i)\right], \tag{78}$$

where the summation symbol sigma has now crept into our formula. Applied to the tandoor oven, this same reasoning would require us to measure the average temperature of the oven at the times $t_1, t_2, \ldots, t_n$; we'd get

$$T_{\text{avg}} = \frac{T(t_1) + T(t_2) + \cdots + T(t_n)}{n}, \quad \text{or} \quad T_{\text{avg}} = \left[\sum_{i=1}^{n} T(t_i)\right]\frac{1}{n}. \tag{79}$$

But there's one thing I've neglected in my analysis. For my tandoori chicken to come out nice and tasty, I'd like the oven to maintain a 500° F average temperature *throughout the time it takes to cook the dish.* It was about 7:15 when I put in the order and it shouldn't take more than a half hour to cook. Thus the measurement times t_i should all lie in the interval $7{:}15 \leq t_i \leq 7{:}45$.

With this in mind let's generalize (79) to the interval $a \leq t_i \leq b$, where $a < b$. We can do this easily as follows:

$$T_{\text{avg}} = \frac{1}{b-a} \left[\sum_{i=1}^{n} T(t_i) \right] \left(\frac{b-a}{n} \right). \tag{80}$$

This sum should be familiar from Chapter 6; it's another example of a Riemann sum. Therefore, we expect that a definite integral will creep in here somewhere. Here's how. In an ideal world, to *completely* ensure that my tandoori chicken is cooking at an average of $500°$ F, I'd be taking measurements every *nanosecond*. Aside from making me the most annoying diner in the history of civilization, this would be highly impractical. Fortunately in Chapter 6 we ran into a similar problem: that of adding up the area of an infinite number of rectangles. Our solution was to instead add up a finite number n and then take the limit as $n \to \infty$. The same reasoning applies here; in this case n represents that total number of temperature measurements. Therefore, the average temperature T_{avg} in the oven between the interval $a \leq t \leq b$ is given by the formula

$$T_{\text{avg}} = \frac{1}{b-a} \lim_{n \to \infty} \left[\sum_{i=1}^{n} T(t_i) \right] \left(\frac{b-a}{n} \right) = \frac{1}{b-a} \int_{a}^{b} T(t)\, dt. \tag{81}$$

This formula tells us that we can calculate the average temperature inside the oven by integrating the temperature function $T(t)$ and then dividing by the length $b - a$ of the time interval!

With about 15 minutes to kill until my dinner arrives I start telling Zoraida about this "integral average" business and its far-reaching consequences. "For example, the temperature of the restaurant itself feels just right, which is controlled by another thermostat," I explain. "This time it's inside the air-conditioning system, but still the same mathematics the tandoor oven uses is at work inside the AC system." My enthusiasm about the topic is not, however, shared by Zoraida.

I can see her interest starting to fade. "Really, any calculation of an average value of a continuous function makes use of the same mathematics," I continue. "Whether it's the average rainfall in a given month, the average number of products a company sells in a quarter, or even the average number of crimes reported in a state, the integral formula for the average helps us compute these quantities." At this point Zoraida is taking a few too many sips of her water; I've lost eye contact. "It's fascinating really, especially since this integral can be calculated by finding areas!" I've barely managed to finish the sentence when the server comes back to bring us our samosas; saved by the food! As we chow down on the appetizer, I'm sure somewhere, at some point, Zoraida realized that there are downsides to marrying a mathematician.

Finding the Best Seat in the House

I managed to keep the conversation nonmathematical for the rest of our dinner, and our Indian feast was indeed tasty. It's now about 8:15, and we start walking over to the theater. On the way over I feel like I'm biting my tongue; there are *tons* of examples of interesting mathematics I see. For instance, the occasional wind gusts remind me of all the neat things fluids can do, and the changing frequency of sound coming from the cars passing us on the street is an example of the Doppler effect. But this time around I keep my mathematical thoughts to myself and just have a "normal" conversation.

We get inside the theater and walk over to screen number 7. We pass the concession stand on the way, where I'm briefly reminded again of how expensive going to the movies has become. Inside theater 7 we're confronted with the age-old problem that every moviegoer faces: where should we sit?

If my earlier rambling about the tandoor temperature revealed one downside of marrying a mathematician, this'll be an example of one upside. As Zoraida stands there, her chin up in the air as she looks around for good seats, I lean over and say "I got this." Like Russell Crowe in *A Beautiful Mind*, formulas start racing through my head.

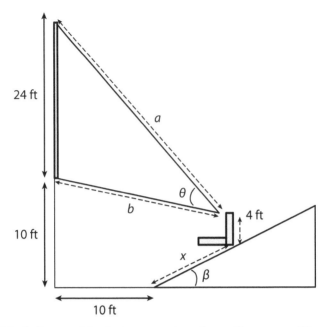

24 ft

10 ft

4 ft

a

θ

b

x

β

10 ft

Figure 7.2. A diagram of the theater parameters, along with someone sitting x feet up the rows in the theater whose eyes are 4 feet off the ground.

In mere *seconds* I crunch the numbers and point to a couple of seats in the third row. "Those are the best seats in the house."

Of course, this didn't happen at all. The truth is that I've done the calculations before, and since the dimensions of the theater haven't changed since then, neither has the answer. By estimating the theater's parameters—like the screen height, the number of rows, and the angle of the stadium seating—I came up with the diagram shown in Figure 7.2. Let me show you how I used this diagram to help us find the best place to sit.[27]

The first step is to quantify what we mean by the "best" view. One way to express this mathematically is to look for the maximum viewing angle θ; in whatever row of the theater this occurs you'll see the totality of the picture most clearly. By using trigonometry, we can determine that a, b, θ, and x are related by the formula[*1]

$$\theta(x) = \arccos\left(\frac{a^2 + b^2 - 576}{2ab}\right), \qquad (82)$$

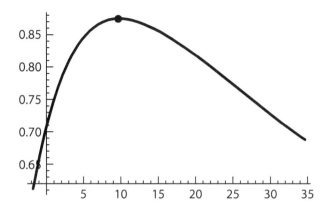

Figure 7.3. The graph of $\theta(x)$, along with its maximum value (the dot on the graph).

where the lengths a and b are given by

$$a^2 = (10 + x \cos \beta)^2 + (30 - x \sin \beta)^2,$$
$$b^2 = (10 + x \cos \beta)^2 + (6 - x \sin \beta)^2. \tag{83}$$

Here the angle β gives the incline angle of the seats, and my estimates put this at about $20°$. Although we could follow our prescription from Chapter 5 and find the stationary points of $\theta(x)$, finding its derivative would be a horror show of its own. Instead, Figure 7.3 shows the plot of the function $\theta(x)$ for $0 \le x \le 35$.

As we see, the maximum value of θ occurs at $x \approx 7.37$. Since the rows of the theater are about 3 feet apart, this means that Zoraida and I should sit somewhere between the second and third rows; this is where I got my original suggestion from. Unfortunately, I don't think my less-than-Oscar-worthy performance was enough to convince her that I worked this out on the spot. Nonetheless, my sly suggestion is one of the *many* upsides of marrying a mathematician.

So how do integrals creep in here? Well, as we sit down to watch the previews we notice that people coming in can now only pick from seats with increasingly suboptimal viewing angles. If the theater company wants to give its guests a relatively happy moviegoing experience, it would be wise to think about providing all of the customers with the best possible viewing angles. One way to do this is to design the seating so that the *average* viewing angle is always *at least* a certain value, say A. If the movie theater has a total of 30 rows—representing an x-interval of $0 \leq x \leq 90$ at 3 feet per row—then this condition can be expressed as

$$\frac{1}{90 - 0} \int_0^{90} \theta(x)\, dx \geq A. \tag{84}$$

I may have eyeballed the theater's parameters, but a construction company could use computer models to adjust the values of a, b, etc., so that this minimum-average-angle condition is satisfied. This is a great example of the dream team in action.

Now, you may be thinking that this idea of using derivatives and integrals in concert for theater construction may be a bit of an overkill. But this isn't as far fetched as it sounds. In fact, a similar analysis goes into building symphony halls. For example, the Boston Symphony Hall was itself built in consultation with the Harvard physicist Wallace Sabine. His expertise helped the hall become one of the world's top three concert halls in terms of acoustics.[28] I don't know if such a concerted effort was made when the theater we're in was designed, but since the movie's about to start, it's about time I stop thinking and just sit back and enjoy the show.

Keeping the T Running with Calculus

It's now about 10:30. Outside the theater there's a lot of foot traffic; dressed-up people are rushing by in both directions. Some look like they're going dancing, and we briefly consider the idea of dancing salsa tonight. It's a lot of fun, but also takes a lot of energy; scrap that. Others look like they're heading to a bar for some drinks. Better option, but add the train ride back and we're looking at getting home around midnight; guess not. Standing there talking this out, Zoraida and I realize that we're actually pretty tired. She looks across the street at the Boylston station, and I nod my head. A few minutes later we're on the D-line train back home.

On the way back I tell Zoraida about that astonishing 1.8 million miles that the MBTA system logged in 2009 (discussed in Chapter 6). "I believe it," she says; "in the mornings I can never find a seat." That leads me to wonder: how much of the 1.8 million miles can be attributed to the D-line that we're on right now? This is a variant of the train maintenance problem we discussed in Chapter 6, since it also involves finding the distance that a particular train has traveled. And based on our earlier work, we can anticipate that definite integrals will appear.

One approach is to know the round-trip distance the D line travels from end to end; multiplying this by the total number of trips it makes in the year would give me an answer. The squeaking sounds of the train's wheels on the track as it rounds a curve point out one problem with this approach: the train tracks follow a *curved* path (see the T map in Figure 7.4(a)). If I had one of those rolling tape measure thingies that surveyors use, I could, theoretically, measure the track's length and then use this multiplication approach. But I don't. Plus, that'd take *forever*, and as we've said before, there *has to* be an easier way.

The answer to our question hinges on our ability to find the length of a curve. Let's call the D-line curve in Figure 7.4(a) $f(x)$, and plot it on a coordinate system (Figure 7.4(b)). Suppose that we zoom in to a segment of this curve, with our "window" having a width Δx and a height Δy (Figure 7.5). The hypotenuse Δz of the triangle in the figure is given by the Pythagorean Theorem:

$$(\Delta z)^2 = (\Delta x)^2 + (\Delta y)^2, \quad \text{or} \quad \Delta z = \sqrt{(\Delta x)^2 + (\Delta y)^2}. \quad (85)$$

By using our old friend the Mean Value Theorem, we can rewrite this quantity as[*2]

$$\Delta z = \sqrt{1 + [f'(x_i)]^2}\,\Delta x, \quad (86)$$

where x_i is inside the interval Δx. If we now imagine splitting the graph into n such segments and making this same approximation in each segment, we'd arrive at the following estimate for the length l of the

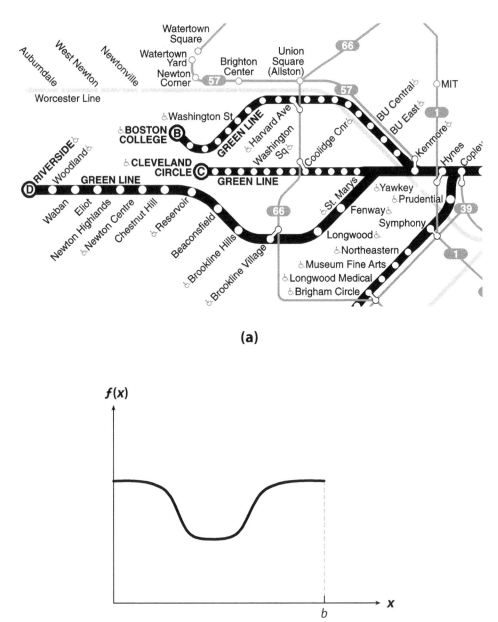

(a)

(b)

Figure 7.4. (a) The MBTA's D-line map showing the stops between the Boylston Street stop and our stop in Newton. (b) The same tracks considered as a function $f(x)$.

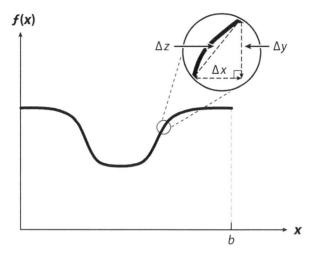

Figure 7.5. A close-up of a segment of the graph of $f(x)$. The length Δz here is the hypotenuse of the triangle whose base and height are Δx and Δy, respectively, and the x-value b represents the eastward distance the Boylston Street station is from our home.

curve:

$$l \approx \Delta z_1 + \Delta z_2 + \cdots + \Delta z_n = \sqrt{1 + [f'(x_1)]^2}\,\Delta x + \cdots$$

$$+\sqrt{1 + [f'(x_n)]^2}\,\Delta x = \sum_{i=1}^{n} \sqrt{1 + [f'(x_i)]^2}\,\Delta x. \qquad (87)$$

If you recognize this as a Riemann sum, then you're on the right track. If we now let the width Δx of these triangles get infinitesimally small by taking the limit as $n \to \infty$, we get

$$l = \lim_{n \to \infty} \sum_{i=1}^{n} \sqrt{1 + [f'(x_i)]^2}\,\Delta x = \int_{0}^{b} \sqrt{1 + [f'(x)]^2}\,dx. \qquad (88)$$

Applied to our case—we live about eight miles west of the Boylston Street stop—to find the distance the train travels in taking us home I'd need to calculate the integral

$$\int_{0}^{8} \sqrt{1 + [f'(x)]^2}\,dx. \qquad (89)$$

Although I don't know the function $f(x)$, as we discussed in Chapter 6 this integral is the area under the function $\sqrt{1 + [f'(x)]^2}$. We could approximate l to any desired accuracy by using the rectangle methods we discussed in the last chapter.[xxvii] Today computers can do this very quickly, so there's really no need for me to go any further.

The formula (88) we've derived can, as usual, be applied to many more situations. For example, manufacturers of furniture, vehicles, or planes use it to help them determine how much material they'll need, since the surfaces of those items tend to be curved and therefore their dimensions are not calculable using simple multiplication.

Look Up to Look Back in Time

The conductor has now announced that our stop is next. This time there was no delay, and we arrive at our stop just a few minutes past eleven; after a short couple-of-minutes walk we'll be home, ready to turn in for the night.

On clear nights like this the entire sky is visible with the naked eye. We start talking about how this is one of the nice things about living a short train ride from the city; there are no skyscrapers and bright lights to wash out the stars. The moon and even some planets are also visible.[xxviii] It all looks so pretty. But this picturesque night sky actually hides some of the deepest secrets of the universe.

One of the fascinating things that I learned as a kid—which no doubt further fueled my interest in math and science—is that every time we look up at the sky we're actually looking back in time. The reason is that even the nearest star (excluding our Sun), Proxima Centauri, is about a whopping 25 *trillion* miles away. These distances are so large that astronomers typically measure them in *light-years*. Proxima Centauri, for example, is about 4.2 light-years away. This means that light emanating from that star takes about 4.2 years to reach us. And

---

[xxvii]Technically, we'd also have to approximate $f'(x)$ from the graph in Figure 7.4(b), but since it's not too crazy-looking this is somewhat feasible.
[xxviii]You can tell the planets from the stars by comparing the bright disks of the former to the pointlike dots of the latter.

here's the mind-bending part: every time you look up at the sky and spot Proxima Centauri, what you're really seeing is the light the star emitted *more than four years ago*. So you're not seeing the star as it is *now*, but as it was more than four years ago!

Okay, okay, so who cares about a star 25 trillion miles away? But would you believe me if I told you that the same reasoning tells us that every sunrise or sunset you've ever seen was a lie? "What?!" you might say. Well, consider the fact that our Sun is about eight "light-minutes" away, meaning that the sun's light takes about eight minutes to reach us. Put another way, this says that the light we *currently* enjoy from our Sun left that star eight minutes ago. Now imagine that some evil empire with some Death Star device straight out of *Star Wars* came and destroyed our Sun. These facts suggest that we wouldn't find out for another eight minutes. What's more, if they instantaneously appeared—using some sort of "warp drive" technology—we wouldn't even *see* them for another eight minutes! Not even the CIA can pull off stunts like these.

As if this wasn't spooky enough, let me just point out the obvious: Proxima Centauri and the Sun aren't the only two stars we can see in the sky. And for each star in the sky that we can see, since each is a *different* number of light-years away, you see a *different* past when you look at each star. Look at the Sun and you're looking eight *minutes* into the past; look at Proxima Centauri and you're looking 4.2 *years* into the past. This idea that "the past is relative" should remind you of the time travel phenomena we discussed in Chapter 3, where we discussed Einstein's results on the relativity of time. There we were talking about traveling into the *future*. Before I turn in for the night let me tie this back to calculus and tell you one last story that highlights perhaps the *ultimate* application of the dream team, differentiation and integration.

The Ultimate Fate of the Universe

The year was 1915, and a young Albert Einstein had just published his Theory of General Relativity. Almost 230 years after the well-known Isaac Newton described the force of gravity through his Universal Law

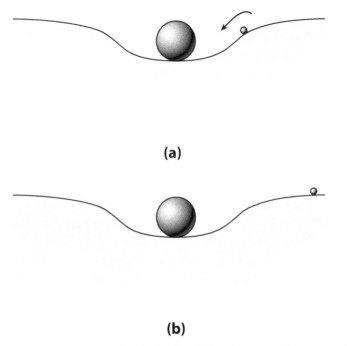

(a)

(b)

Figure 7.6. (a) A pea is attracted to the bowling ball at the center of a mattress due to the way the ball curves the mattress. (b) If the same pea is too far away to "feel" the curvature caused by the ball, it is no longer attracted to the ball.

of Gravitation, this relatively unknown scientist claimed that Newton was wrong about gravity. For Dr. Newton, the force of gravity depended on how close together two masses were. However, Newton's theory contained a troubling implication: for two masses separated by the most unimaginably large distance, moving one would *instantaneously* affect the gravitational force the other felt. For Einstein, who in 1905 had discovered the speed of light to be the cosmic speed limit, the notion of instantaneous effects didn't sit well. So, being the genius that he was, he came up with a radical alternative to Newton's theory: gravity is actually the curving of space by matter.

To wrap our heads around this idea, imagine placing a bowling ball in the middle of a mattress (Figure 7.6). Naturally, the region closest to the ball curves more than the region farther away from it. Now take a pea and place it somewhere on the mattress; one of two things will happen depending on where it's placed relative to the ball. Put it sufficiently close to the ball and let it go, and it'll fall inward toward the ball (Figure 7.6(a)); however, place it far enough away and it'll just stay put (Figure 7.6(b)). This thought experiment shows that what "attracts" the pea to the ball is not some spooky instantaneous force. Instead, it's the *curvature* of the mattress that is responsible for the attraction we would see were the pea placed close enough to the ball. As this statement makes clear, this "gravitational force" still depends on the distance between the ball and the pea. But unlike Newton's theory, this view of gravity solves the "instantaneous" problem: remove the bowling ball and it'll take some time for the mattress's springs to communicate this change in curvature to the pea. These "gravitational waves" are one of the hallmarks of Einstein's theory (predicted by him a year later in 1916). They tell us that any changes in the distances of massive objects *do not* result in instantaneous changes in the gravitational force the objects feel. Instead, these waves have to propagate first—at the speed of light—before they can communicate the change in the force of gravity.

In the course of two years—the period between 1915 and 1916— Albert Einstein managed not only to prove Newton wrong, but also to replace his theory of gravity with a much more correct one. Newton's passive gravity was now understood as the *consequence* of objects falling into the "valleys" created by massive bodies' curving of the very fabric of space and time. Moreover, the discovery that Einstein's equations *propagated* gravity via gravitational waves resolved the instantaneous communication.

So what does this have to do with calculus? The easy answer is that Einstein's equations are written in terms of *derivatives*, but solving them requires *integration* (cue the dream team)! Some of the solutions can be used to study the entire universe, and the predictions you can make are truly astonishing. Let me tell you that story.

In 1927, shortly after Einstein published his equations, the Belgian astronomer Georges Lemaitre used them to make the surprising prediction that the universe was expanding. What's more, he even estimated the rate of expansion. A short two years later the American astronomer Edwin Hubble confirmed this. He gathered observational data on how fast galaxies were receding away from us and found the

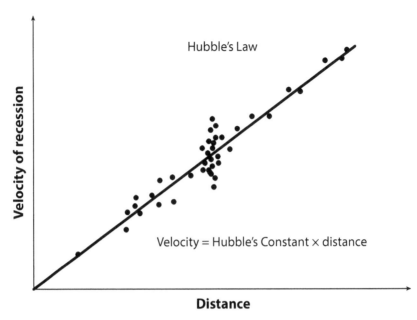

Figure 7.7. A plot of the velocity of distant galaxies against their distance from Earth (each dot represents a distant galaxy). The slope of the line shown represents the Hubble constant H_0. Image from http://imagine.gsfc.nasa.gov/YBA/M31-Velocity/hubble-more.html.

simple relationship

$$v = H_0 d, \tag{90}$$

where v is the galaxy's velocity, d is its distance from Earth, and H_0 is called the *Hubble constant* (see Figure 7.7).

However, despite its name, the Hubble constant is not constant throughout time. Instead, $H(t)$ satisfies the equation

$$H'(t) = -H^2(1 + q), \tag{91}$$

where q is called the *deceleration parameter*. The "deceleration" here reflects scientists' long-held belief that although the universe was expanding, it was slowing down in its expansion. With so much matter in the universe, scientists reasoned, everything would eventually be attracted to something closer to the center. Like an underwater explosion, the universe would eventually cave in on itself, resulting in what some would call a "Big Crunch." In the context of calculus, an object (even if it's a galaxy) that is moving away from you at a slowing rate has a negative acceleration, or a *positive* deceleration (q-parameter).

In 1998 the three astrophysicists Saul Perlmutter, Brian Schmidt, and Adam Reiss *shocked* us all with the discovery that the universe is *not* decelerating; instead, they found it to be *accelerating* (giving a *negative* deceleration parameter).[xxix] This changes the prognosis for our universe. Instead of suggesting a Big Crunch, it suggests that at some *very* distant future time, everything in the universe will be *very* far away from everything else. This rosier picture is nonetheless equally saddening, since it suggests that planetary systems will eventually become isolated from each other (even more than they are now).

With the help of derivatives, the $H'(t)$ equation has now yielded some astonishing information about the fate of our universe. But what about its beginnings? If the universe is expanding—like a balloon does when we blow air into it—then if we imagine going backwards in time, everything we see in the universe must have, at one point, come from a single *very dense* blob of matter and energy. It's from this state that scientists think the universe was born, in what we today call a Big Bang. The natural question is: how long ago did this happen? If I haven't foreshadowed it enough already, the answer will involve an integral (think "dream team").

The Age of the Universe

The possible fates of the universe are numerically described by the values of a number Ω known as the *density parameter*. A universe with $\Omega > 1$ would eventually cave in on itself, suffering the Big Crunch we discussed earlier. A universe with $\Omega < 1$ would expand forever, much like we believe today our universe will. Using this, a simplified formula

---

[xxix] These scientists shared the Nobel Prize in Physics in 2011 for their discovery.

for the age T of the universe is[29]

$$T = \frac{1}{H_0} \lim_{z \to \infty} \int_0^z \frac{dz}{(1+z)\sqrt{\Omega\left[(1+z)^3 - 1\right] + 1}}. \qquad (92)$$

If this expression looks scary, just think about the steps involved. As we learned in Chapter 6, to calculate the integral above we'd first like to have an antiderivative $F(z)$. The Fundamental Theorem of Calculus would then tell us that the answer to the integral is $F(z) - F(0)$. Finally, we'd find the limit as $z \to \infty$ of this expression and divide by Hubble's constant to arrive at a formula for T. After integrating and taking the limit, the formula for T becomes[30]

$$T = \frac{2}{3H_0} \frac{1}{\sqrt{1-\Omega}} \ln\left[\frac{1+\sqrt{1-\Omega}}{\sqrt{\Omega}}\right]. \qquad (93)$$

This formula tells us that all we need is an Ω value to find T. Using our best current estimates for Ω, we get an age of about 13.75 billion years.[31]

So there you have it; in just a few pages we've used calculus to contemplate the ultimate fate of the universe *and* to estimate its age. The mere fact that there is a body of knowledge that even allows us to think about such things continues to astound me. And we did it all by using the same two pillars of calculus that we've been discussing all throughout this book: differentiation and integration.

Now consider this. We've only focused on the mathematics of *calculus* in this book. Imagine how much more *the rest* of mathematics—geometry, topology, abstract algebra, and more—can teach us. I'm instantly reminded of the medieval scientists we discussed in Chapter 1, to whom just figuring out that motion on Earth is parabolic was a big deal. Picture their faces if they were told that we can calculate the age of the universe.

Back at home, I'm finally in bed and ready to get to sleep. The lights are off and the day is ending. I'm not thinking about getting 7.5 hours of sleep since tomorrow is Saturday; instead, I'm thinking of all the mathematics I've seen throughout the day. From the theoretical developments to the practical applications, it's all been right under our noses the whole time. As I come to the end of my reflection, I close my eyes with a smirk; I realize that ending this book with a discussion of the Big Bang theory gives me the right to make one awesome claim: I've gone out with the Biggest Bang of them all.

IF YOU'VE MADE IT THIS FAR, let me be the first to say that I'm proud of you. As I mentioned in the preface, many of us are unfortunately intimidated by mathematics, thinking that it's either too abstract or too difficult to understand. I hope that what you've read in this book has helped you see that you're already familiar with a lot of mathematics; the only prerequisite is curiosity. Next time you have coffee, take some time to notice how the temperature cools, or study the patterns that the milk you put in it makes as you swirl it around. Or next time you feel a gust of wind, look around for some leaves and you'll likely see a vortex in the making.

With that said, I'd now like to give you the "takeaways" from each chapter to help you spread the good word about calculus.

Chapter 1: Functions are the building blocks of mathematics, and they can be found *everywhere.*

Chapter 2: Derivatives describe change, and thus wherever there is change derivatives can be found.

Chapter 3: "Mathematizing" a problem often helps us understand it better.

Chapter 4: Calculus, and mathematics in general, connects seemingly unrelated phenomena.

Chapter 5: Through the mathematics of optimization, calculus helps us make life better.

Chapter 6: Integrals undo derivatives, and whenever quantities need to be added integrals are not far behind.

Chapter 7: Analyzing a problem with both derivatives and integrals— the dream team—can lead to profound insights.

I'd also like to make what I often tell my students is a "public service announcement." Many of the examples I presented made drastic simplifications, yet we know that reality is rarely ever simple. However, building on simple assumptions has historically been one of the main strengths of experimental science. Aristotle thought that objects, when dropped, fell toward Earth because it was "natural" for them to be on the ground. Galileo wondered how long it took objects to fall, and conducted enough experiments to mathematically describe how objects fall toward Earth. Then Newton took this a step further and described *all motion* through his three laws. This "Legos principle" is at the heart of the success of modern science.

The Legos principle also works well in mathematics, but there's one crucial difference with experimental science: mathematics is eternal. Even ancient mathematics—like the Pythagorean Theorem—is never disproven. The fact that $a^2 + b^2 = c^2$ for planar right triangles is true means that it will remain true forever. But how, then, is progress made? Often what we mathematicians do is to change the assumptions. For example, what if we look at triangles not on a plane but instead on a curved surface (like that of a sphere)? Then the Pythagorean Theorem is no longer valid. Instead, new and interesting *non-Euclidean* geometry results. Or take the simple fact that $12 + 1 = 13$. What if we now insisted that $12 + 1 = 1$? This may seem crazy, but wait until noon tomorrow and then ask someone for the time one hour later. She'll likely say one in the afternoon, i.e., that $12 + 1 = 1$.[xxx]

You may never have noticed this peculiar thing about how we tell time, but that's exactly the main message of this book: look close enough at the world around you and you'll find mathematics everywhere, connecting things you never thought were related in a beautiful and often deep way. And even if you need to make some simple assumptions about what you see, changing those assumptions often leads to even more interesting mathematics. That's part of what makes mathematics fun, and I highly encourage you to continue exploring everything the subject has to offer.

Oscar Edward Fernandez
Newton, MA[xxxi]

[xxx]If you ask people in the military, they might still say 13, but to them $24 + 1 = 1$.
[xxxi]P.S. I live in Newton and I wrote a book about calculus—how cool is that!

APPENDIX A

||||||||||||||||||||||||||||||||||||||||||||||||||||||||||||||||||||||||||||||||||||

FUNCTIONS AND GRAPHS

EVEN THOUGH YOU MIGHT NOT REALIZE IT, functions are all around you. The temperature outside your home is a function of time; the cost of filling up at the gas station is a function of how many gallons you pump into your car's gas tank; and the amount of calories you burn doing an exercise is a function of how long you've been exercising. Mathematically, we call the inputs—time, gallons pumped, and exercise duration in our examples—the *independent variables* and use the letter x to denote them. We call the outputs—temperature, gas cost, and calories burned—the *dependent variables* and use the letter y to denote them. We write $y = f(x)$ to express the fact that the dependent variable y depends on, or *is a function of*, the independent variable x.

One important feature that our three examples have in common is that each input leads to a *unique* output. This means, for example, that after having exercised for 30 minutes you couldn't have burned 100 calories *and* 120 calories; it has to be one or the other but not both. This is the main idea behind what we mean mathematically by a function: a function is a relationship between a collection of inputs and a collection of outputs where each input is assigned to a unique output. We call the collection of inputs the *domain* of the function and the collection of outputs the *range* of the function.

A very useful way to visualize a function is to graph it. Typically we plot the independent variable x on a horizontal axis and the dependent variable y on the vertical axis. For example, for our gas cost function if we assume that 1 gallon of gas costs \$4 and we pump x gallons, then

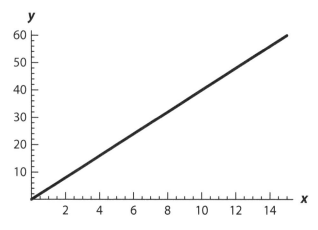

Figure A.1. The graph of the function $f(x) = 4x$.

the total cost $y = f(x) = 4x$. The graph of this function is given in Figure A.1.

In the terminology above, the domain of $f(x)$ is any positive number along with zero, since we could pump as many gallons as we'd like, but it doesn't make sense to pump, for example, -3 gallons. The range of $f(x)$ is also any positive number along with zero. Moreover, we can *evaluate* $f(x)$ at a point in its domain to obtain a point in its range (for example, $f(2) = 4(2) = 8$, meaning that the cost of pumping 2 gallons is $8).

The function graphed in Figure A.1 belongs to the class of *linear* functions $g(x) = mx + b$, so named since the graphs of all linear functions are lines. The numbers m and b here have important meanings. Since $g(0) = m(0) + b = b$, this tells us that b is the y-value at $x = 0$. For our function $f(x)$ from Figure A.1 we see that $b = 0$, so that the point $x = 0$ and $y = 0$, denoted by the pair $(0, 0)$, is on the graph of $f(x)$. We will call b the *y-intercept*, since it's the y-axis point the graph of $g(x)$ touches when it crosses the y-axis. The second value m is called the *slope*. Mathematically, the slope of a linear function like $g(x)$ is defined by

$$m = \frac{g(b) - g(a)}{b - a},$$

(94)

for any two nonequal x-values a and b. We can spot the slope of the function $f(x)$ from Figure A.1 right away from its equation: $f(x)$ has slope 4. Great, but what does this mean? Well, if we choose the x values $x = 0$ and $x = 1$, then equation (94) says that $4(1 - 0) = f(1) - f(0)$. This equation tells us that when the x-values change by one unit the y-values change by four units. This is often expressed by saying that "as x runs from zero to one, y rises by four." Accordingly, slope is often referred to as the "rise over run."

The next thing to notice about the graph in Figure A.1 is that it doesn't "double-back" on itself. This is a general feature of graphs of functions that results directly from our definition, since any doubling-back would result in two y-values (the outputs) for a given x-value (the input). This conclusion gives us a way to determine whether a graph we're looking at is the graph of a function: a graph isn't a function if at least one vertical line intersects the graph more than once. This is known as the *vertical line test*.

One example of a graph that violates the vertical line test is the equation

$$(x^2 + y^2 - 1)^3 - x^2 y^3 = 0.$$

As Figure A.2 shows, the graph of this equation fails the vertical line test in many places. Too bad, I really had my heart set on this one.

There are many types of functions, but the most common ones are:

1. **Power Functions**. These are functions of the form $f(x) = ax^n$, for a, n numbers. Examples: $f(x) = 2x^2$, $f(x) = \frac{1}{3}x^{3/2}$.
2. **Polynomials**. When you restrict n to only take on the values $0, 1, 2, \ldots$, and add the resulting power functions you get a function of the form $f(x) = a_0 + a_1 x + a_2 x^2 + \ldots + a_n x^n$. These are the polynomial functions. They are often named by the highest power of x that occurs. For example, the polynomial $f(x) = 1 + 2x$ is a *linear* function, while the polynomial $f(x) = 4 + 3x - 7x^2$ is a *quadratic* function.
3. **Rational Functions**. We can divide two polynomial functions together to get a rational function. Example: $f(x) = (1 + 2x + 3x^2)/(3 - x)$. You might notice a bit of a problem with this

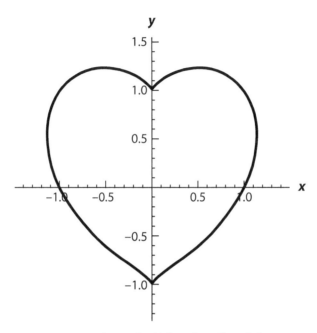

Figure A.2. The graph of $(x^2 + y^2 - 1)^3 - x^2 y^3 = 0$.

example: if you try to *evaluate* the function at $x = 3$, you get $f(3) = 34/0$. Dividing by zero is a problem. When we divide two nonzero numbers we get a unique answer (such as $18/9 = 2$), from which we can express the numerator uniquely as the quotient multiplied by the divisor (continuing, $18 = 2(9)$). The problem with zero as a divisor is that the quotient is not unique. For example, $0 = 2(0)$ but also $0 = 7(0)$. For this reason, we forbid division by zero in all of our formulas. This discussion tells us that the domain of $f(x)$ *excludes* $x = 3$.

4. **Trigonometric Functions**. The most freguently used trig functions are the sine function $f(x) = \sin x$ and the cosine function $g(x) = \cos x$. These functions are *periodic*, meaning that their y-values—and consequently their graphs—repeat. Referring to the graph in Figure A.3(c) (a sine function), we can see that the y-value $y = 0$ vertically cuts the graph in half. We call this y-value the *midline*, and denote it by C. We also see that the

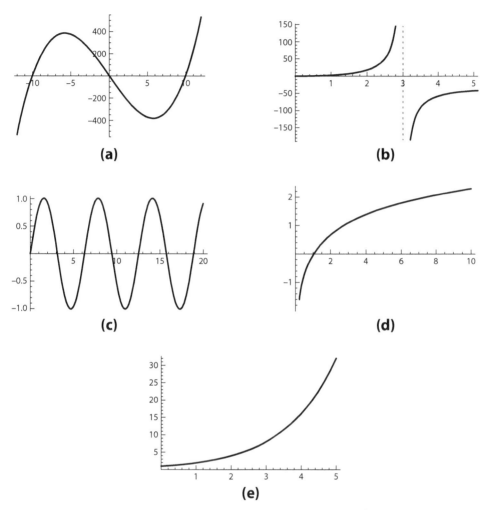

Figure A.3. Graphs of (a) the polynomial $f(x) = 1 - 100x + x^3$, (b) the rational function $f(x) = (1 + 2x + 3x^2)/(3 - x)$, (c) the trig function $f(x) = \sin x$, (d) the logarithmic function $f(x) = \ln x$, and (e) the exponential function $f(x) = 2^x$. Note that the vertical line shown in graph (b) is *not* part of the graph of the function. The function graphed in (b) is the one from the "rational functions" discussion above. The vertical line at $x = 3$ is reflecting the fact that the function is undefined at $x = 3$. It is an example of a vertical asymptote.

largest y-value is $y = 1$. The difference between the maximum value and the midline is called the *amplitude A*; $A = 1$ for our example. The last important number associated with sine and cosine functions is the *frequency* F. This number tells us how many cycles[xxxii] there are in an interval of unit length. An associated concept is that of the *angular frequency*, denoted by B. This number tells us how many cycles there are in an interval of length 2π, where $\pi \approx 3.14$. The frequency and angular frequency are related by $F = B/2\pi$. The last important number is the *period* $T = 2\pi/B = 1/F$, which tells us how long an x-interval is needed for the graph of $f(x)$ to complete one full cycle. These numbers can be used to construct a sine or cosine function: $f(x) = A\sin(Bx) + C$, or $g(x) = A\cos(Bx) + C$. The period of the sine function in Figure A.3(c) is $T = 2\pi$, its angular frequency is $B = 1$, and its frequency is $F = 1/2\pi$. And using the A, C, T numbers for Figure A.3(c) we can say that $f(x) = 1 \cdot \sin(x) + 0$ is the function graphed in the figure.

5. **Exponential Functions**. These are functions of the form $f(x) = ab^x$. The number a is the *initial value*, since $f(0) = a$, and the number b is called the *base*. In this book we consider only exponential functions whose base $b > 0$. Among the set of possible bases, we often use the number $e \approx 2.71$ as the base. Examples: $f(x) = 2e^x$, $g(x) = -7(2^x)$. Two important rules that exponential functions obey are (1) $a^x b^x = (ab)^x$ and (2) $a^x a^y = a^{x+y}$.

6. **Logarithmic Functions**. These functions have the form $f(x) = a\log_b x$. Here we again restrict to $b > 0$, and b is also called the *base* of the logarithm. The two most frequently used bases are $b = 10$, in which case we simply write $\log x$ instead of $\log_{10} x$, and base $b = e$, in which case we usually write $\ln x$ instead of $\log_e x$. Logarithmic and exponential functions are inverses of each other. This means that, for example, if $y = 5^x$ then $x = \log_5 y$.

[xxxii]A "cycle" is the portion of the graph between two peaks, or equivalently between two troughs.

1. We can translate the information given about the typical sleep cycle into values for A, B, C for the function $f(t) = A\cos(Bt)+C$. From Appendix A we know that $B = 2\pi/T$, so that from $T = 1.5$ we get $B = (4/3)\pi$. Since the highest sleep stage is 0 and the lowest is -4, the midline is the middle value $C = -2$. The amplitude A is then the maximum value minus the midline: $A = 0 - (-2) = 2$. Using these values in our function above leads to the $f(t)$ equation given in the chapter.

2. We can simplify the equation $f(t) = -1$ to

$$2\cos\left(\frac{4\pi}{3}t\right) = 1, \qquad \text{or} \qquad \cos\left(\frac{4\pi}{3}t\right) = \frac{1}{2}.$$

By taking the inverse cosine of both sides we get

$$\frac{4\pi}{3}t = \frac{\pi}{3}, \frac{5\pi}{3}, \frac{7\pi}{3}, \frac{11\pi}{3}, \frac{13\pi}{3}, \ldots,$$

where the negative values have been omitted since we're interpreting t as time. Solving for t yields $t = 0.25, 1.25, 1.75, 2.75, 3.25, \ldots$. With these numbers in mind, we can picture the t-values for which $f(t) \geq -1$ as the intervals between the gray dots in Figure A1.1 that contain peaks of $f(t)$.

3. We can rearrange $L = 20\log_{10}(50{,}000\,p)$ to read $\log_{10}(50{,}000\,p) = L/20$. Using the fact that $10^{\log_{10}x} = x$ for $x > 0$, we then have

$$50{,}000\,p = 10^{L/20}, \qquad \text{or} \qquad p(L) = \frac{1}{50{,}000}10^{L/20}.$$

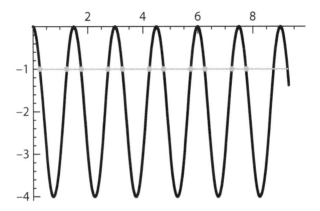

Figure A1.1. The intersection of the line $y(t) = -1$ with the graph of $f(t)$.

4. Intuitively, we know that objects that are accelerating have speeds that change with time (think of an airplane accelerating from rest to takeoff). If we measure the accelerating object's speed at times t_a and t_b and get $v(t_a)$ and $v(t_b)$, then we say that the object's acceleration a over that time interval was

$$a = \frac{v(t_b) - v(t_a)}{t_b - t_a}.$$

For our water molecule with acceleration $a = -g$ we are measuring its speed over the time interval $[0, t]$. Therefore,

$$-g = \frac{v(t) - v_y}{t - 0}, \qquad \text{or} \qquad v(t) = v_y - gt.$$

I should note that this derivation depends on the fact that a is constant.

5. From $x(t) = v_x t$ we have that $t = x/v_x$. Substituting this into $y(t) = 6.5 + v_y t - (g/2)t^2$ yields

$$y(x) = 6.5 + \frac{v_y}{v_x}x - \frac{g}{2v_x^2}x^2.$$

1. Using formula (2) from Chapter 2, the average rate of change of AAPL over the past 12 months is

$$\frac{P(12) - P(0)}{12 - 0} = \frac{\$610.76 - \$390}{12\,\text{months}} \approx 18.4 \ \$/\text{month},$$

while the AROC over the past 4 months is

$$\frac{P(12) - P(8)}{12 - 8} = \frac{\$610.76 - \$625}{4\,\text{months}} = -3.56 \ \$/\text{month}.$$

Note that the units of the AROC (in this case $/month) are the units of the numerator (in this case $) divided by the units of the denominator (in this case months).

2. Since our interval is now $t = 8$ to $t = 8 + h$, in the notation of formula (2) it follows that $a = 8$ and $b = 8 + h$. Therefore, using this in formula (2) yields

$$m_{\text{avg}} = \frac{P(8 + h) - P(8)}{8 + h - 8} = \frac{P(8 + h) - P(8)}{h},$$

which verifies formula (3) of Chapter 2.

3. According to our definition of the derivative at the point $t = a$ from equation (4), we have that

$$T'(0) = \lim_{h \to 0} \frac{T(0 + h) - T(0)}{h} = \lim_{h \to 0} \frac{75 + 85e^{-0.318h} - 160}{h}$$

$$= \lim_{h \to 0} \frac{85(e^{-0.318h} - 1)}{h}.$$

1. The larger claim here is that if a function $f(x)$ is increasing, then its derivative $f'(x)$ is positive. Let's give some justification of this by recalling the definition of the derivative from equation (4) of Chapter 2:

$$f'(x) = \lim_{h \to 0} \frac{f(x+h) - f(x)}{h}.$$

Now, a function is increasing if when you increase the x-values the y-values increase. In this context, this means that if $h > 0$, then $f(x+h) - f(x) > 0$. Therefore, if $h > 0$, then both the numerator and the denominator are positive. A similar argument shows that if $f(x)$ is decreasing, then $f'(x)$ is negative.

While you're here, I might as well tell you that if $f'(x)$ is positive, then $f(x)$ is increasing. Some justification for this is provided by the approximation

$$f'(x) \approx \frac{f(x+h) - f(x)}{h},$$

which is accurate when h is very small. From this we can reason that since the left-hand side is positive, the right-hand side *should* be as well. This turns out to be true. Although this isn't a formal proof—that is not the goal of this book—you can repeat the same argument to convince yourself that if $f'(x)$ is negative, then $f(x)$ is decreasing.

2. Let's see how we can solve equation (12) for the terminal velocity of the falling water droplet. Starting from that equation, which is

$$(m(t)v(t))' = 32m(t),$$

we use the product rule for derivatives to differentiate the left-hand side. We obtain

$$m'(t)v(t) + m(t)v'(t) = 32m(t).$$

But remember that through equation (10) we related the rate of change of $m(t)$ to $m(t)$ itself. Using this relationship we obtain

$$2.3m(t)v(t) + m(t)v'(t) = 32m(t).$$

Since every term in this equation contains $m(t)$, dividing the entire equation by $m(t)$ (which is "legal" since $m(t)$ is never zero) will eliminate all the $m(t)$'s and result in

$$2.3v(t) + v'(t) = 32.$$

Now, normally to solve an equation like this we'd use the methods of differential equations (a more advanced course than calculus), but since we're sticking to calculus we'll just have to make do with what we've got. To start making progress, let's try to make the right-hand side zero (that always helps). We do this by introducing the new function $z(t) = v(t) - (32/2.3)$. Then $v(t) = z(t) + (32/2.3)$, and $v'(t) = z'(t)$. Substituting these in we arrive at

$$32 + 2.3z(t) + z'(t) = 32, \quad \text{or, equivalently,} \quad 2.3z(t) + z'(t) = 0$$

$$\text{or} \quad z'(t) = -2.3z(t).$$

Let's now think carefully about what this last equation is telling us. Quite literally, the equation says that whatever $z(t)$ is, it's proportional to its own derivative. We know of only one function that has this property: e^{at}. It's derivative is—by the chain rule—ae^{at}, a function proportional to the original e^{at}. Therefore, let's set $z(t) = e^{at}$. Then $z'(t) = ae^{at}$, and the last equation reads

$$ae^{at} = -2.3e^{at}.$$

We can divide through by e^{at} since it's never zero, and so obtain $a = -2.3$. Thus $z(t) = e^{-2.3t}$, and so $v(t) = e^{-2.3t} + 32/2.3$. This is a perfectly fine solution, except that at $t = 0$ it tells us that the initial velocity $v(0) = 32/2.3$. We'd like to enforce $v(0) = 0$ since we're assuming the raindrop falls from rest. Luckily, we can fix this easily by rewriting $v(t)$ as $v(t) = ke^{-2.3t} + 32/2.3$, where k is yet to be determined. This is *still* a solution of our equations (as you can check), but now $v(0) = k + 32/2.3$. Since we want $v(0) = 0$, this tells us that $k = -32/2.3$. Finally, our solution is

$$v(t) = -\frac{32}{2.3}e^{-2.3t} + \frac{32}{2.3} = \frac{32}{2.3}(1 - e^{-2.3t}).$$

This reproduces equation (13) of Chapter 3.

3. It's clear from Figure 3.1 that as t gets larger $v(t)$ approaches $32/2.3$. This wording should remind you of our computations of limits in Chapter 2. Indeed, to show that $v(t)$ cannot exceed $32/2.3$ ft/s as t gets larger and larger what we want to confirm is that

$$\lim_{t \to \infty} v(t) = \frac{32}{2.3}.$$

We could make a limit table as we've done in the past, but we'll just reason this one out. Remember that

$$e^{-2.3t} = \frac{1}{e^{2.3t}}$$

by the laws of exponents. So $v(t)$ can be rewritten as

$$v(t) = \frac{32}{2.3} - \frac{32}{2.3e^{2.3t}}.$$

From here we can see clearly that as t gets larger and larger $e^{2.3t}$ gets larger and larger, so that everything after the minus sign is getting closer and closer to zero. Therefore, as $t \to \infty$, the only thing that survives is $32/2.3$.

4. The definition of the derivative from equation (4) of Chapter 2 was

$$f'(a) = \lim_{h \to 0} \frac{f(a+h) - f(a)}{h}.$$

If we now change variables by writing $h = x - a$, then as $h \to 0$ we have $x - a \to 0$, or $x \to a$. Substituting gives

$$f'(a) = \lim_{x \to a} \frac{f(x) - f(a)}{x - a}.$$

5. To find the equation of the line tangent to the graph of $f(x)$ at the point $(a, f(a))$, we use the point-slope formula:

$$y - y_0 = m(x - x_0).$$

We know that this is a *tangent* line, so that its slope is the derivative at $x = a$: $m = f'(a)$. And since we also know that it passes through $(a, f(a))$, then $x_0 = a$ and $y_0 = f(a)$. Using all this we get

$$y - f(a) = f'(a)(x - a), \qquad \text{or} \qquad y = f(a) + f'(a)(x - a).$$

6. We first need to calculate $J'(x)$ for $J(x) = 3,000/\pi x^2$. Rewriting this as $J(x) = (3,000/\pi)x^{-2}$ and using the power rule—which states that if $f(x) = x^n$, then $f'(x) = nx^{n-1}$—we get

$$J'(x) = \frac{3,000}{\pi}(-2x^{-3}) = -\frac{6,000}{\pi x^3}.$$

Using this, the approximation in (15) becomes

$$J(6) - J(5) \approx J'(5)(6 - 5) =$$

$$-\frac{6,000}{\pi(8,046.72)^3}(1,609.34) \approx -5.9 \times 10^{-6} \text{W/m}^2,$$

where we've used the conversion 5 miles $= 8,046.72$ meters.

7. Using the power rule it follows that $f'(x) = 2x$ and $f''(x) = 2$, while $g'(x) = -2x$ and $g''(x) = 2$. By plugging in zero we obtain the values in equations (16).

8. Let's begin by rewriting the equation as

$$z(x) = y(1 - x)^{-1/2},$$

where $x = v^2/c^2$. We can now use the linear approximation

$$z(x) \approx z(0) + z'(0)(x - 0).$$

By the chain rule we have

$$z'(x) = \frac{y}{2}(1 - x)^{-3/2}, \qquad \text{and so} \qquad z'(0) = \frac{y}{2}.$$

Therefore, our approximation gives

$$z(x) \approx y + \frac{y}{2}x = y\left(1 + \frac{1}{2}x\right) = y\left(1 + \frac{v^2}{2c^2}\right).$$

1. To differentiate a quotient $f(x)/g(x)$ (provided $g(x) \neq 0$) we use the *quotient rule*:

$$\left(\frac{f(x)}{g(x)}\right)' = \frac{f'(x)g(x) - f(x)g'(x)}{(g(x))^2}.$$

Applying this to our $A(x)$ function yields

$$A'(x) = \frac{p'(x)x - p(x)(1)}{x^2} = \frac{xp'(x) - p(x)}{x^2}.$$

2. Since the denominator of $A'(x)$ is never negative, it follows that $A'(x) > 0$ only when the numerator is positive. So

$$xp'(x) - p(x) > 0, \qquad \text{or} \qquad p'(x) > \frac{p(x)}{x} = A(x).$$

3. Starting from

$$p'(x) > A(x) = \frac{p(x)}{x}, \qquad \text{we have that} \qquad \frac{p'(x)}{p(x)} > \frac{1}{x}. \quad \text{(A1.1)}$$

Now note that functions $p(x)$ satisfying

$$\frac{p'(x)}{p(x)} = \frac{k}{x},$$

where $k > 1$, will automatically satisfy condition (A1.1). To solve this equation, we use the fact that the derivative on the left-hand side is the *logarithmic derivative* of the function $p(x)$, since

$$(\ln p(x))' = \frac{p'(x)}{p(x)}.$$

Therefore, we have that

$$(\ln p(x))' = \left(\ln Cx^k\right)',$$

where $C > 0$, from which it follows that $p(x) = Cx^k$.

4. We can rewrite the equation for $I(t)$ as

$$I(t) = 20(1 + 3e^{-20kt})^{-1}.$$

Using the chain rule,

$$(f(g(x)))' = f'(g(x))g'(x),$$

it follows that

$$I'(t) = -20(1 + 3e^{-20kt})^{-2}(-60ke^{-20kt})$$

$$= k\frac{20}{1 + 3e^{-20kt}} \cdot \frac{60e^{-20kt}}{1 + 3e^{-20kt}}$$

$$= kI \cdot \frac{20 \cdot 3e^{-20kt}}{1 + 3e^{-20kt}} = kI\left(20 - \frac{20}{1 + 3e^{-20kt}}\right) = kI(20 - I).$$

5. Let's make a plot of I' versus I. From equation (25) it follows that what we see is a quadratic function (Figure A4.1). Since the "x-intercepts" are $I = 0$ and $I = 20$, the maximum of the function is halfway between these two, at $I = 10$. We see from the graph that before $I = 10$ the function is increasing (and the tangent lines therefore have positive slopes), and after $I = 10$ the function is decreasing (and the tangent lines therefore have negative slopes). But these tangent lines are the derivatives of I', otherwise known as I''. Therefore, before $I = 10$ we have $I''(t) > 0$, and afterward we have $I''(t) < 0$, making $C = 10$ the "y-value" of the point of

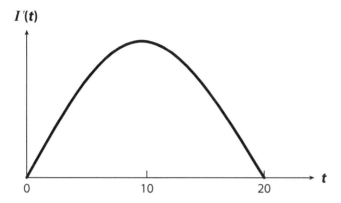

Figure A4.1. The graph of the function $20kI - kI^2$.

inflection. The time t^* at which this happens can be obtained from solving the equation $I(t^*) = 10$ for t^*.

6. In mathematics, "eventually" is usually a synonym for $t \to \infty$. So by using equation (26) we have that

$$\lim_{t \to \infty} \frac{20}{1 + 3e^{-20kt}} = \frac{20}{1 + 3 \lim_{t \to \infty} e^{-20kt}} = 20.$$

7. Calculating the limit gives

$$\lim_{t \to \infty} \frac{(a-c)p_0}{bp_0 + ((a-c) - bp_0)e^{-(a-c)t}}$$

$$= \frac{(a-c)p_0}{bp_0 + ((a-c) - bp_0) \lim_{t \to \infty} e^{-(a-c)t}} = \frac{(a-c)p_0}{bp_0} = \frac{a-c}{b},$$

where we've assumed that $c < a$, so that $e^{-(a-c)t} \to 0$ as $t \to \infty$.

8. With

$$B(t) = \left(B(0) + \frac{100s}{r} \right) e^{rt/100} - \frac{100s}{r},$$

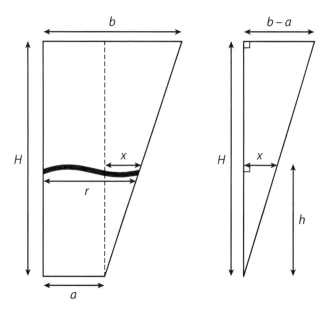

Figure A4.2. A cross section of the cup.

the chain rule gives

$$B'(t) = \left(B(0) + \frac{100s}{r}\right)\left(\frac{r}{100}\right)e^{rt/100} = \frac{r}{100}\left(B(t) + \frac{100s}{r}\right)$$

$$= \frac{r}{100}B(t) + s.$$

9. Over that 20-year period, your total contributions would amount to $20 \times \$5{,}000 = \$100{,}000$. Subtracting this, along with the initial $\$30{,}000$ from $B(20)$, yields $\$220{,}280.31$, which is 68.78% of the $\$320{,}280.31$ that the account gained over the 20-year period.

10. Figure A4.2 shows a profile of the cup and the liquid.

    If we break up the liquid's radius r into $r = a + x$, then by similar triangles we have

$$\frac{x}{h} = \frac{b-a}{H}, \quad \text{or} \quad x = \frac{(b-a)h}{H}, \quad \text{and} \quad r = a + \frac{(b-a)h}{H}.$$

Substituting this representation of r into the frustum volume equation we arrive at

$$V = \frac{\pi h}{3} \left[\left(a + \frac{(b-a)h}{H} \right)^2 + a \left(a + \frac{(b-a)h}{H} \right) + a^2 \right],$$

which simplifies to

$$V = \frac{\pi}{3} \left(3a^2 h + \frac{3a(b-a)}{H} h^2 + \frac{(b-a)^2}{H^2} h^3 \right).$$

11. To differentiate $V(h(t))$ with respect to t we use the chain rule. We get $V'(h(t))h'(t)$, where $V'(h(t))$ is just the derivative of the function $V(h)$ with respect to h. Since

$$V'(h) = \frac{\pi}{3} \left(3a^2 + \frac{6a(b-a)}{H} h + \frac{3(b-a)^2}{H^2} h^2 \right),$$

all that's left to do is to multiply by $h'(t)$. The result is precisely formula (34) in Chapter 4.

1. From $f(r) = kr^4$ we know that $f'(r) = 4kr^3$. Using this in $dr = f'(a) \, dr$ gives

$$dr = 4ka^3 \, dr.$$

2. The mathematical statement that guarantees this is called *Fermat's Theorem*. The version relevant to our purposes states that if a function $f(x)$ is differentiable at some x_0 in the interval $a < x < b$ (meaning that $f'(x_0)$ exists) and $f'(x_0) \neq 0$, then x_0 is not an extremum of f. So, all points of a differentiable function—a function whose derivative $f'(x)$ exists for every value of x—that are not stationary points cannot be extrema. But since Fermat's Theorem says nothing about the endpoints a, b, we also need to consider these when searching for the extrema of f.

3. The path from point A in Figure A5.1 through the branching point B and onto the endpoint C can be divided into two segments of lengths l_1 and l_2.

   The total distance l the blood travels is $l = l_1 + l_2$. Poiseuille's second law then tells us that the total resistance encountered is

$$R = c \left(\frac{l_1}{r_1^4} + \frac{l_2}{r_2^4} \right),$$

which means that we need to find l_1, l_2. From the triangle portion of the figure we see that

$$\sin \theta = \frac{M}{l_2}, \quad \text{or} \quad l_2 = \frac{M}{\sin \theta} = M \csc \theta.$$

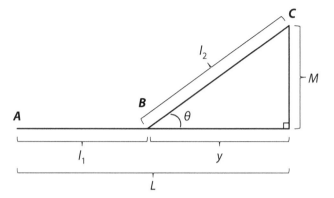

Figure A5.1. The two paths, ABC and AB, that blood would take down the artery.

Now, since the full length of the main vessel is L, if we call the portion that forms the base of the triangle y, then $L = l_1 + y$. From this we see that $l_1 = L - y$, so we need to know what y is. From the triangle we can determine y:

$$\tan \theta = \frac{M}{y}, \quad \text{or} \quad y = \frac{M}{\tan \theta} = M \cot \theta.$$

Therefore, $l_1 = L - M \cot \theta$. Using this in the equation for the resistance R gives

$$R = c \left(\frac{L - M \cot \theta}{r_1^4} + \frac{M \csc \theta}{r_2^4} \right).$$

4. The derivative of $R'(\theta)$ is

$$R'(\theta) = c \left[\frac{M}{r_1^4} \csc^2 \theta - \frac{M \csc \theta \cot \theta}{r_2^4} \right].$$

Setting this to zero yields

$$\frac{M}{r_1^4} \csc^2 \theta = \frac{M \csc \theta \cot \theta}{r_2^4}, \quad \text{or} \quad \frac{r_2^4}{r_1^4} = \frac{M \csc \theta \cot \theta}{M \csc^2 \theta} = \cos \theta.$$

5. With $R(x) = 12{,}000 + 140x - 2x^2$, we use the power rule to obtain

$$R'(x) = 140 - 4x.$$

6. Substituting in $x = 35$ into $R(x)$ gives

$$R(35) = 12{,}000 + 140(35) - 2(35)^2 = \$14{,}450.$$

7. Using the right triangle in Figure 5.5 we know that

$$y^2 = (6 - x)^2 + (2.1)^2.$$

Substituting this into $g = \dfrac{x}{36} + \dfrac{y}{29}$ gives

$$g(x) = \frac{x}{36} + \frac{\sqrt{(6 - x)^2 + 4.41}}{29}.$$

8. Let's first rewrite $g(x)$ as

$$g(x) = \frac{x}{36} + \frac{1}{29}\left[(6 - x)^2 + 4.41\right]^{1/2}.$$

Using $g(x)$ from above we have

$$g'(x) = \frac{1}{36} + \frac{1}{29}\left(\frac{1}{2}[(6 - x)^2 + 4.41]^{-1/2}(-2(6 - x))\right)$$

$$= \frac{1}{36} - \frac{6 - x}{29\sqrt{(6 - x)^2 + 4.41}}.$$

Setting this equal to zero yields

$$\frac{1}{36} = \frac{6 - x}{29\sqrt{(6 - x)^2 + 4.41}}.$$

By cross-multiplying and squaring both sides this simplifies to

$$841 \left[(6 - x)^2 + 4.41\right] = 1{,}296(6 - x)^2,$$

and by distributing and combining like terms we arrive at

$$455x^2 - 5{,}460x + 1{,}2671.2 = 0.$$

We can find the roots of this quadratic equation by using the quadratic formula, and we get $x \approx 3.14$ and $x \approx 8.86$. However, since only $x \approx 3.14$ is inside the interval $0 \leq x \leq 6$, we reject the second solution $x \approx 8.86$.

1. The two areas we need to add are the area of a rectangle, which is $A = bh$ (where b and h are the base and height of the rectangle, respectively), and the area of a triangle, which is $A = (1/2)bh$ (where b and h are the base and height of the triangle, respectively). We have

$$A_I + A_{II} = (35)(0.0042) + \frac{1}{2}(35)(0.0083 - 0.0042) \approx 0.22 \text{ mile.}$$

2. Adding the left-hand sides of all the equations gives

$$\frac{b}{n}v(t_0) + \frac{b}{n}v(t_1) + \cdots + \frac{b}{n}v(t_{n-1}) = [v(t_0) + v(t_1) + \cdots + v(t_{n-1})]\frac{b}{n}.$$

Adding the right-hand sides of the equations gives

$$\left[s\left(\frac{b}{n}\right) - s(0)\right] + \left[s\left(\frac{2b}{n}\right) - s\left(\frac{b}{n}\right)\right] + \cdots$$
$$+ \left[s(b) - s\left(\frac{(n-1)b}{n}\right)\right] = s(b) - s(0).$$

Comparing the two gives

$$[v(t_0) + v(t_1) + \cdots + v(t_{n-1})]\frac{b}{n} = s(b) - s(0),$$

and writing the left-hand side as a Riemann sum gives

$$\sum_{i=0}^{n-1} v(t_i)\frac{b}{n} = s(b) - s(0).$$

3. From the power rule we know that

$$\left(\frac{x^{n+1}}{n+1}\right)' = x^n,$$

provided that $n \neq -1$ is a number. Therefore, it follows that

$$\frac{x^{n+1}}{n+1} = \int x^n \, dx \qquad (n \neq -1).$$

Now, the one slightly confusing thing is that this isn't the only antiderivative of x^n. The functions

$$\frac{x^{n+1}}{n+1} + 1 \qquad \text{or} \qquad \frac{x^{n+1}}{n+1} + 14$$

would also do, since their derivatives all produce x^n. So the most general antiderivative of x^n is

$$\int x^n \, dx = \frac{x^{n+1}}{n+1} + C \qquad (n \neq -1),$$

where C is referred to as an *arbitrary constant*. Applying this to find the antiderivative of $-g$ yields

$$v(t) = \int -g \, dt = -gt + C.$$

When $t = 0$ this gives $v(0) = C$, meaning that C is the initial velocity. To reflect this we now denote C by v_0, so that

$$v(t) = v_0 - gt.$$

4. Recall that the derivative of a sum of two functions is the sum of the derivatives of the two functions. The same is true for integration:

$$y(t) = \int (v_0 - gt) \, dt = \int v_0 \, dt + \int -gt \, dt = y_0 + v_0 t - \frac{1}{2}gt^2.$$

5. The integral

$$1 - \int_0^5 \frac{1}{5} e^{-t/5} \, dt$$

can be calculated by using the method of u-substitution. If we introduce the variable $u = -t/5$, then the differential $du = -1/5 \, dt$, or $dt = -5 \, du$. Under this substitution, the limit of integration $t = 0$ becomes $u = -(0)/5 = 0$, and the limit of integration $t = 5$ becomes $u = -5/5 = -1$. Using all of this information we get

$$1 - \int_0^{-1} \frac{1}{5} e^u (-5 \, du) = 1 + \int_0^{-1} e^u \, du = 1 - \int_{-1}^0 e^u \, du.$$

We can now use the Fundamental Theorem of Calculus; since $(e^x)' = e^x$, the antiderivative of e^u is just e^u, and so

$$1 - \int_{-1}^0 e^u \, du = 1 - (e^0 - e^{-1}) = e^{-1} \approx 0.368.$$

1. Applying the law of cosines to the triangle in Figure 7.2 yields

$$(24)^2 = a^2 + b^2 - 2ab \cos\theta, \quad \text{or} \quad 2ab \cos\theta = a^2 + b^2 - 576.$$

Solving this for θ yields

$$\cos\theta = \frac{a^2 + b^2 - 576}{2ab}, \quad \text{or} \quad \theta = \arccos\left(\frac{a^2 + b^2 - 576}{2ab}\right).$$

To determine a and b, we can split up the triangle in Figure 7.2 into two right triangles (see Figure A7.1). Both of these triangles have base

$$z = 10 + x \cos\beta,$$

and using the Pythagorean theorem we get

$$a^2 = (10 + x \cos\beta)^2 + (34 - 4 - x \sin\beta)^2,$$
$$b^2 = (10 + x \cos\beta)^2 + (10 - 4 - x \sin\beta)^2,$$

which simplifies to the a and b formulas in Chapter 7.

2. Since the quantity Δy in the expression

$$\Delta z = \sqrt{(\Delta x)^2 + (\Delta y)^2}$$

represents the change in $y = f(x)$ over the interval Δx, assuming that f is a differentiable function (which the function in

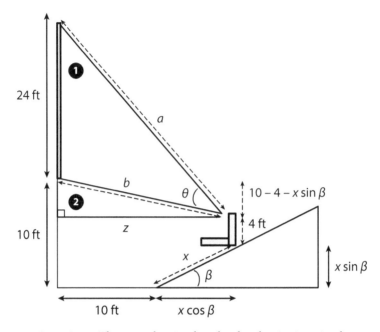

24 ft

10 ft

10 ft

Figure A7.1. The two right triangles related to the viewing triangle.

Figure 7.5 is), the Mean Value Theorem tells us that

$$\Delta y = f'(x_i)\Delta x,$$

for some x_i in the interval Δx. Using this we get

$$\Delta z = \sqrt{(\Delta x)^2 + [f'(x_i)]^2(\Delta x)^2} = \sqrt{1 + [f'(x_i)]^2}\,\Delta x.$$

1. Klein, S. and Thorne, B.M. *Biological Psychology*. New York: Worth Publishers, 2007.
2. Klein and Thorne, *Biological Psychology*.
3. Carr, N. *The Big Switch: Rewiring the World, from Edison to Google*. New York: W. W. Norton & Company, 2008.
4. This "war of currents" is discussed at length in the excellent book *AC/DC: The Savage Tale of the First Standards War*, by Tom McNichol (San Francisco: Jossey-Bass, 2006).
5. *WBUR Highlights & History*. n.d. Retrieved from http://www.wbur.org/about/highlights-and-history.
6. Avison, J. *The World of Physics*. Cheltenham, UK: Thomas Nelson and Sons, 1989.
7. Gelfand, S.A. *Essentials of Audiology*. New York: Thieme Medical Publishers, 2009.
8. Gelfand, *Essentials of Audiology*.
9. Avison, *The World of Physics*.
10. Galileo's life and accomplishments have been the subject of many books. A recent excellent account is David Wootton's *Galileo: Watcher of the Skies* (New Haven: Yale University Press, 2010).
11. An excellent account of the incremental scientific progress in medieval times that helped Galileo make his discoveries can be found in *God's Philosophers*, by James Hannam (London: Icon, 2009).
12. Kornblatt, S. *Brain Fitness for Women*. San Francisco: Red Wheel/Weiser, 2012.
13. Downs, A. *Still Stuck in Traffic: Coping with Peak-Hour Traffic Congestion*. Washington, D.C.: Brookings Institution Press, 2004.
14. "The American Commuter Spends 38 Hours a Year Stuck in Traffic." *The Atlantic*, February 6, 2013. Retrieved from http://www.theatlantic.com/business/archive/2013/02/the-american-commuter-spends-38-hours-a-year-stuck-in-traffic/272905/.
15. "The American Commuter."
16. U.S. Department of Commerce. *Population Estimates*. n.d. Retrieved from http://www.census.gov/popest/data/historical/.
17. Got, J. Richard III. *Time Travel in Einstein's Universe: The Physical Possibilities of Travel through Time*. New York: Mariner Books, 2002.
18. Landa, Heinan. "You vs. Your Inbox." *Washington Business Journal.*, January 23, 2013. Retrieved from http://www.bizjournals.com/washington/blog/techflash/2013/01/you-vs-your-inbox-guest-blog.html.

19. *The Shocking Cost of Internal Email Spam*. n.d. Retrieved from http://www.vialect.com/cost-of-internal-email-spam.
20. Mui, Ylan Q. "Americans Saw Wealth Plummet 40 Percent from 2007 to 2010, Federal Reserve Says." *Washington Post*, June 11, 2012. Retrieved from http://articles.washingtonpost.com/2012-06-11/business/35461572_1_median-balance-median-income-families.
21. Shell, Adam. "Holding Stocks for 20 Years Can Turn Bad Returns to Good." *USA Today*, June 8, 2011. Retrieved from http://usatoday30.usatoday.com/money/perfi/stocks/2011-06-08-stocks-long-term-investing_n.htm.
22. This is an old problem; a detailed treatment of it dates back to at least 1926, when Cecil D. Murray studied it in his article titled "The Physiological Principle of Minimum Work Applied to the Angle of Branching of Arteries," published in *The Journal of General Physiology* (in 1926).
23. Wilson, Susan. *Boston Sites and Insights: An Essential Guide to Historic Landmarks in and around Boston*. Boston: Beacon Press, 2004.
24. American Public Transportation Association. *2011 Public Transportation Fact Book*. Washington, D.C.: American Public Transportation Association, 2011.
25. Blakemore, Judith E. Owen, Berenbaum, Sheri A., and Liben, Lynn S. *Gender Development*. New York: Psychology Press, 2008.
26. Wilson, *Boston Sites and Insights*.
27. This question was studied by Kevin Mitchell in "Calculus in a Movie Theater," *UMAP Journal* 14(2) 1993, 113–135.
28. Boston Symphony Orchestra. *Acoustics*. n.d. Retrieved from http://www.bso.org/brands/bso/about-us/historyarchives/acoustics.aspx.
29. See "Distance Measures in Cosmology," by David W. Hogg, available at the Cornell University Library online archive: arXiv:astro-ph/9905116.
30. See Kevin Krisciunas's article titled "Look Back Time, the Age of the Universe, and the Case for a Positive Cosmological Constant," available at Cornell University Library online archive: arXiv:astro-ph/9306002v1.
31. NASA. *WMAP—Age of the Universe*. n.d. Retrieved from http://map.gsfc.nasa.gov/universe/uni_age.html

Lightning Source UK Ltd.
Milton Keynes UK
UKHW021829061222
413289UK00005B/195

9 780691 157559